NUCLEAR POWER

A Reference Handbook

Other Titles in ABC-CLIO's
Contemporary
World Issues
Series

Agricultural Crisis in America, Dana L. Hoag
Biodiversity, Anne Becher
Endangered Species, Clifford J. Sherry
Environmental Justice, David E. Newton
Genetic Engineering, Harry LeVine III
Indoor Pollution, E. Willard Miller and Ruby M. Miller
Natural Disasters: Floods, E. Willard Miller and Ruby M. Miller
Natural Disasters: Hurricanes, Patrick J. Fitzpatrick
The Ozone Dilemma, David E. Newton
Recycling in America, Second Edition, Debi Strong

Books in the Contemporary World Issues series address vital issues in today's society such as genetic engineering, pollution, and biodiversity. Written by professional writers, scholars, and nonacademic experts, these books are authoritative, clearly written, up-to-date, and objective. They provide a good starting point for research by high school and college students, scholars, and general readers as well as by legislators, businesspeople, activists, and others.

Each book, carefully organized and easy to use, contains an overview of the subject, a detailed chronology, biographical sketches, facts and data and/or documents and other primary-source material, a directory of organizations and agencies, annotated lists of print and nonprint resources, and an index.

Readers of books in the Contemporary World Issues series will find the information they need in order to have a better understanding of the social, political, environmental, and economic issues facing the world today.

NUCLEAR POWER

A Reference Handbook

Harry Henderson

CONTEMPORARY
WORLD ISSUES

ABC-CLIO

Santa Barbara, California
Denver, Colorado
Oxford, England

Library of Congress Cataloging-in-Publication Data
Henderson, Harry, 1951–
Nuclear power : a reference handbook / Harry Henderson.
p. cm.—(Contemporary world issues)
Includes index.
ISBN 1-57607-128-6 (hardcover : acid-free paper)
1. Nuclear engineering. 2. Nuclear engineering—History.
3. Nuclear accidents. 4. Nuclear energy—Government policy.
I. Title. II. Series.

TK9146 .H45 2000
333.792'4'09—dc21

00-010255
CIP

06 05 04 03 02 01 00 10 9 8 7 6 5 4 3 2 1

ABC-CLIO, Inc.
130 Cremona Drive, P.O. Box 1911
Santa Barbara, California 93116-1911

This book is printed on acid-free paper ♾.
Manufactured in the United States of America.

Contents

Preface, xi

1 Introduction, 1
The Discovery of Nuclear Energy, 1
Nuclear Fission, 3
The First Nuclear Reactor, 4
Nuclear Power and Its Problems, 5
Nuclear Power Comes On-Line, 6
Radiation, Health, and the Environment, 7
The Search for Safety, 9
Three Mile Island, 12
The Chernobyl Disaster, 14
The Nuclear Waste Problem, 17
Yucca Mountain, 19
Transporting Nuclear Waste, 20
Nuclear Proliferation, 22
What Do You Do with an Old Nuke? 23
A Nuclear Sunset? 25
A Nuclear Renaissance? 27
Conclusion: An Uncertain Future, 28
References, 29

2 Chronology, 31

3 Biographical Sketches, 45
Bernard Baruch (1870–1965), 45
Hans Albrecht Bethe (1906–), 46
Niels Bohr (1885–1962), 47
Helen B. Caldicott (1938–), 47

Jimmy Carter (1924–), 48
Bernard Cohen (1924–), 48
Barry Commoner (1917–), 49
Dwight D. Eisenhower (1890–1969), 49
Enrico Fermi (1901–1954), 50
John Gofman (1918–), 50
Bourke B. Hickenlooper (1896–1971), 51
Henry M. Jackson (1912–1983), 51
David E. Lilienthal (1899–1981), 52
Lise Meitner (1878–1968), 52
G. C. Minor (1937–1999), 53
Mayumi Oda (1941–), 53
J. Robert Oppenheimer (1904–1967), 54
Dixy Lee Ray (1914–1994), 55
Hyman G. Rickover (1900–1987), 55
Glenn T. Seaborg (1912–1999), 56
Karen Silkwood (1946–1974), 56
Lewis L. Strauss (1896–1974), 57
Grace Thorpe (1924–), 57
Stewart Udall (1920–), 58

4 Facts, Illustrations, and Documents, 59
Design and Operation of Nuclear Power Plants, 59
Basic Nuclear Plant Designs, 59
Nuclear Contribution to the U.S. Energy Supply, 62
Reactors in the United States, 63
World Nuclear Reactors, 77
The Fuel Cycle and Nuclear Waste, 79
The Nuclear Fuel Cycle, 79
Locations of Spent Fuel Storage, 85
The Yucca Mountain Proposal, 85
Yucca Mountain Nuclear Waste Depository, 85
Opposition to Nuclear Power, 102
Nuclear Power's Failed Promise, 102
The Nuclear Waste Controversy, 105
What's Wrong with Burying Nuclear Waste at Yucca Mountain? 106
Why We Call It "Mobile Chernobyl," 113
Legal Aspects of Nuclear Power, 116
Regulations, 116
International Cooperation and Treaties, 119

Regulatory Agencies, 121
Court Cases, 122

5 Directory of Organizations, 131

6 Print Resources, 147
References and Overviews, 148
Bibliographies, 148
General Reference and Overview, 153
Advocacy, 156
Antinuclear, 156
Pronuclear, 158
Regulations, Politics, and Policymaking, 159
Regulatory Policy and Politics, 159
Case Studies, 166
Future Prospects, 167
International Issues, 169
Development of Nuclear Power in Other Nations, 169
Nuclear Proliferation, 172
Nuclear Disasters, 176
Three Mile Island, 176
Chernobyl, 176
Nuclear Waste Issues, 178
Environmental Effects, 178
Waste Disposal and Management, 180
Miscellaneous Issues, 183
Fusion Power, 183
Nuclear Power in Space, 183
Other, 184
Periodicals, 186

7 Nonprint Resources, 191
Websites, 192
General Resources, Databases, and Archives, 192
Historical and Documentary, 194
Regulatory, Legal, and Governmental Information, 195
Technical Information about Nuclear Power, 197
Radiation and Nuclear Waste, 199
Antinuclear, 200
News Groups, 201
Audiovisual Materials, 202

Background and Overviews, 202
Antinuclear Advocacy, 205
Pronuclear Advocacy, 210
Chernobyl and Other Nuclear Accidents, 211
Nuclear Waste and Environmental Effects, 213
Sources for Audiovisual Materials, 215

Glossary, 219
About the Author, 235
Index, 237

Preface

At the turn of the twenty-first century, nuclear power may seem to be old news. The first commercial nuclear reactor went into service in 1957. During the 1950s Americans had a profoundly ambivalent relationship with nuclear energy. They feared the Bomb, dug shelters, and wondered whether the next in history's recurring wars would be the last. At the same time, nuclear power promised abundant, cheap electricity to light and power a booming economy.

Although not living up to the hopes of its most enthusiastic advocates, nuclear power grew rapidly in the 1960s and into the 1970s. But two events threatened that growth by changing the attitudes of many Americans toward this form of energy. First was the growing popular concern about the environment and a variety of threats to it, symbolized by the first Earth Day in 1970. The radiation discharges and wastes generated by nuclear plants became a greater concern as researchers learned more about the long-term effects of radiation on humans and the environment. Second was the occurrence of two highly publicized nuclear accidents: the partial meltdown at the Three Mile Island nuclear plant in Pennsylvania in 1979 and the far more catastrophic disaster at Chernobyl in the Ukraine in 1986.

Even as industrialized but oil-poor nations such as France and Japan enthusiastically committed themselves to nuclear power as a major source for their future energy, planning and construction of new nuclear plants in the United States ground to a halt, and no new plants have been ordered since 1978. Today, dozens of plants are reaching the end of their license period, and a variety of economic factors, including the relatively low prices of alternative fuels (such as coal and gas) make the construction of new nuclear plants in the United States seem to be an unlikely proposition for the foreseeable future.

Nevertheless, there are many reasons why nuclear power and its associated issues will remain very important for decades to come. There remains the legacy of radioactive waste, which has to be disposed of safely and monitored indefinitely. Plants going out of service must be decommissioned, a complex procedure that creates additional waste. Accompanying this nuclear legacy is an ongoing concern about the potential diversion of nuclear materials to terrorists or rogue states that might be able to use them to fashion nuclear weapons. And perhaps most significantly, at a time when there is growing concern about the global warming caused mainly by the burning of fossil fuels such as coal and oil, nuclear energy offers an alternative that does not produce greenhouse gases. The need for this alternative energy source, plus proposed safer reactor designs, may lead to a reconsideration of the role of nuclear power in the new century.

This book provides both the background information needed to understand today's nuclear power issues and a guide to the most useful and accessible resources.

The first chapter gives an overview of developments in nuclear energy from the discovery of radiation and nuclear fission to the role of the U.S. government in promoting nuclear power, the growth of the industry, the accidents and controversies that challenged the viability of nuclear power, and today's pressing issues of nuclear waste disposal and nuclear proliferation.

Chapter 2 provides a chronological view of the events covered in the first chapter and fills in further details in its year-by-year listing. It helps provide historical context, suggesting how technological and regulatory developments may be related to broad changes in American society from the late 1940s to the 1990s.

Chapter 3 presents biographical sketches of some of the key figures involved in the discovery, development, and opposition to nuclear power. They range from scientists and politicians to regulators and activists (many of whom are also scientists, of course). Although other individuals might have been added, the selections reflect diverse viewpoints that are reflective of the many approaches that can be taken to nuclear issues.

Chapter 4 moves from the narrative and anecdotal view to the statistical and graphical presentation of information. Through tables, charts, and maps, it provides an overview of how nuclear reactors work, how much energy they produce, where they are located, and the plans for disposing of nuclear waste. In addition to information from government and industry, statements from antinuclear groups are included to provide a balanced and diverse perspective. Finally, because the legal system is so often involved in nuclear issues, an overview of regulations, regulatory agencies, and court cases is included.

Chapter 5 presents a listing of groups involved with nuclear power issues. They include government regulatory agencies, academic and research institutions, industrial groups, and antinuclear activist groups. Nearly every group has a World Wide Web address listed, and many also have E-mail contacts. This reflects the primary role the Internet now plays both for organizations seeking to communicate information and for researchers and students seeking to find it.

Chapter 6 is devoted to print resources—mainly books. The listings are divided into topical categories to make it easy to find readings on a given aspect or issue. Selections range from popular accounts suitable for high school students and adults, to somewhat more technical works. In addition, some periodicals are also given as a starting point for more specialized research.

In Chapter 7 the power of the Internet is further revealed in a listing of many websites that reflect a wide range of viewpoints. These sites include large collections of information and links to other websites as well as archive and database sites. Many sites reflect specific issues such as nuclear waste and proliferation or provide resources for activists and concerned citizens. A selection of Usenet newsgroups is also provided. These groups can be a good way to keep up with recent developments and to get in touch with other researchers. The chapter closes with a selection of audiovisual (mainly video) products that provide background, documentaries, investigative reports, interviews, and other presentations on nuclear issues.

Chapter 8 provides an extensive glossary of the scientific and technical terms readers are likely to encounter when reading about nuclear matters. The book's index, of course, provides a quick way to find specific names or concepts.

No one book can provide everything a student or researcher needs to study such a complex and multifaceted subject as nuclear power. However, the bibliographical and Web-related resources presented here open unlimited pathways to pursue the subject to whatever degree of detail might be desired, from high school report to doctoral thesis.

1

Introduction

In order to understand the complex issues surrounding nuclear power it is first necessary to understand just what makes the atom unique as a source of energy. People have known about fossil fuels such as coal for hundreds of years, and the use of petroleum has been familiar for more than a century. But nuclear energy was not even an idea until about one hundred years ago. Our overview of nuclear energy therefore begins with the discovery of its nature and implications.

The Discovery of Nuclear Energy

What is nuclear energy, and how did people learn how to use it? Most of the energy we use comes from the chemical process of combustion (burning coal, oil, or gas). By the late nineteenth century scientists understood how combustion and other basic chemical reactions work. Atomic theory, proposed by John Dalton at the start of the nineteenth century, explained that all matter is made of atoms of a number of basic substances, called elements, each of a particular weight, which can combine in various ways to form compounds. The atoms themselves were believed to be indivisible (which is what the word "atom" means in Greek).

But in the mid-1890s a series of startling discoveries began to change this picture radically. In 1895 a German physicist, Wilhelm Roentgen, was experimenting with a Crookes (cathode ray) tube and accidentally discovered a strange form of invisible energy that could pass through solid objects. He called it X-rays, with the X standing for "unknown."

Besides being useful to doctors for setting broken bones, X-rays provided a powerful new tool for physicists to use to probe the structure of matter. In 1896 French physicist Antoine Becquerel was wondering whether X-rays might be emitted by substances that had absorbed sunlight. He chose to investigate an interesting substance called uranium, which had been discovered by German chemist Martin Heinrich Klaproth in 1789. This gray metal, the heaviest of the natural elements, had greatly intrigued him, so much that he declared, "I am certain that the investigation of uranium, starting from its natural resources, will lead to many new discoveries. I confidently recommend that one who is looking for new study topics to give particular consideration to the uranium compounds" (Inam ur Rehman 1993, 77).

Becquerel exposed a compound containing uranium to the sun, then placed it on a photographic plate that had been covered with black paper. Although light could not reach the film, it was exposed anyway. Becquerel at first assumed that the uranium, after absorbing sunlight, had emitted X-rays that passed through the paper to the film. But much to his surprise, the uranium exposed the film even after a series of cloudy, sunless days. The uranium itself was producing energy, but how?

The famous scientific husband-and-wife team of Marie and Pierre Curie painstakingly sifted through uranium ores in search of the source of the mysterious "radioactive" energy. They found that the uranium itself could not account for all the radiation, and they eventually found new sources of radiation: the elements thorium, polonium, and radium. Marie Curie soon realized that "the radioactivity of these substances [is] decidedly an atomic property. It seems to depend on the presence of atoms of the . . . elements in question, and is not influenced by any change of physical state or chemical decomposition" (Grady 1992, 34). In fact, she had discovered a new kind of energy: atomic energy.

But weren't atoms featureless, indivisible balls of primal stuff? How could they spontaneously generate energy? It soon became clear that the interior of the atom was actually a complex world containing smaller particles. In 1897 British physicist Joseph John Thompson discovered a negatively charged particle called the electron that was smaller than an atom. Indeed, the electron turned out to be the part of the atom that defined its chemical characteristics and that was responsible for the flow of energy we call electricity.

But what was at the core, or nucleus, of the atom? A New Zealand–born British physicist named Ernest Rutherford, who started his career as Thompson's assistant in the burgeoning new field of X-ray physics, soon found out. Carefully measuring the radiation coming from uranium, Rutherford discovered that it actually had two components: positively charged particles he called "alpha," and negatively charged "beta" particles—basically, energetic, free-moving electrons. (In 1900 French physicist Paul Villard discovered a kind of radiation that was not a particle but a form of energy like light but with a much shorter wavelength. He called it gamma rays.)

In 1919 Rutherford and his associate Frederick Soddy began to bombard atoms with fast-moving alpha particles. When an alpha particle hit a nitrogen atom, for example, it was sometimes absorbed. The nitrogen atom instantly emitted a positively charged particle that was equivalent to the nucleus of hydrogen, the simplest of the elements. This particle became known as the proton, for the Greek word for "first." Meanwhile, the nitrogen, retaining one proton, had changed to a heavier element, oxygen. Scientists had discovered that atoms could actually be transmuted, or changed from one element to another, by a change in the number of protons. During the 1920s Rutherford and other researchers discovered other nuclear reactions. Rutherford's measurements also made him realize that "the energy latent in the atom must be enormous compared to that rendered free in ordinary chemical change" (Andrade 1964, 72).

Nuclear Fission

Science-fiction writers such as H. G. Wells began to explore the tremendous potential of nuclear energy. But to turn atoms into an energy source, there had to be a way to start and control a steady series of nuclear reactions. The key to controlled nuclear reactions was the neutron, discovered by British physicist James Chadwick in 1932. Physicist Isidor I. Rabi pointed out the neutron's potential use for creating nuclear reactions:

> Since the neutron carries no charge, there is no strong electrical repulsion to prevent its entry into nuclei. In fact, the forces of attraction which hold nuclei together may pull the neutron into the nucleus. When a neutron enters a nucleus, the effects are about as cata-

> strophic as if the moon struck the earth. The nucleus is violently shaken up by the blow, especially if the collision results in the capture of the neutron. A large increase in energy occurs and must be dissipated, and this may happen in a variety of ways, all of them interesting. (Rhodes 1986, 209)

In 1934, Italian physicist Enrico Fermi bombarded uranium with neutrons, resulting in the creation of new radioactive isotopes of elements near uranium on the periodic table. (An isotope of an element is a variety that has a different number of neutrons in the nucleus. For example, all uranium atoms have 92 protons, but they can have different numbers of neutrons, giving atomic weights of 234, 235, or 238. Different isotopes of an element behave the same chemically, but can have different effects in nuclear reactions.)

Just how interesting nuclear reactions could be became apparent only when Austrian physicist Lise Meitner and her nephew Otto Frisch discussed some experiments on a winter's day in 1938. Earlier that year, the German chemists Otto Hahn and Fritz Strassmann had also bombarded uranium with neutrons, but instead of finding that small parts of the atoms had been chipped off to yield nearby elements, they detected the presence of barium, a far lighter element. Meitner and Frisch soon realized that some uranium atoms were not being merely chipped, but split in two. They used the biological term "fission" to describe this process.

The First Nuclear Reactor

As the news about nuclear fission spread, its implications were quickly grasped by the worldwide community of physicists. Nuclear fission changed a small amount of the mass in the target atom into energy, and Albert Einstein's famous equation $E=MC^2$ indicated that the energy released by even that small mass would be tremendous. Further, each fission was also accompanied by the emission of one or more neutrons. If uranium atoms could be arranged so that the neutrons emitted by each fission triggered another fission, the result could be a controlled chain reaction that could produce abundant, cheap energy. But if the atoms were arranged so that each fission's neutrons produced on the average more than one additional fission, the result would be an uncontrolled chain reaction—and an explosion far greater than that achievable by any chemical explosive. Suppose such a weapon fell into the hands of German chancellor Adolf Hitler, who was even then preparing to launch World War II?

When anxious physicists approached Albert Einstein, he wrote to President Franklin D. Roosevelt, warning him of the destructive potential of nuclear energy. Roosevelt responded to Einstein's pleadings, and by 1942 a massive secret effort to build an atomic bomb was under way. To build the bomb, scientists and engineers had to solve all the fundamental problems of nuclear power. They developed techniques to separate the fissionable uranium-235 isotope from the much more abundant uranium-238 by filtering it from the gas uranium hexafluoride. Because they weren't sure if they could get enough fissionable material that way, a separate effort sought to manufacture the fissionable artificial element plutonium-239. That effort required the development of a controlled fission reaction in a nuclear reactor.

Building a reactor would require mastering the design and engineering techniques that would become basic to the nuclear power industry. Under the stands at the University of Chicago football stadium, scientists constructed a simple reactor, or "nuclear pile," consisting of bricklike blocks of graphite in which uranium was embedded. The graphite served as the "moderator," slowing down the fission neutrons enough so they could cause new fissions. On December 2, 1942, the reactor "went critical," which is to say it produced a self-sustaining fission reaction, for the first time. This date can be said to mark the beginning of the nuclear age.

Nuclear Power and Its Problems

The first the general public heard of nuclear energy was when atomic bombs were dropped on the Japanese cities of Hiroshima and Nagasaki in August 1945. Experts interviewed by the press saw a brilliant silver lining in the ominous mushroom clouds: A pellet of uranium the size of a thimble can produce energy equivalent to 1,780 pounds of coal, 149 gallons of oil, or 17,000 cubic feet of natural gas. The public was assured that it was only a matter of time before the atomic cornucopia would create a world in which, according to science writer David Dietz,

> all forms of transportation will be freed at once from the limits now put upon them by the weight of present fuels. . . .
>
> Instead of filling the gasoline tank of your automobile two or three times a week, you will travel for a year on a pellet of atomic energy the size of a vitamin pill. . . . The day is gone when nations will fight for oil. . . .
>
> The world will go permanently off the gold standard once the era of Atomic Energy is in full swing. . . . With the aid of atomic energy the scientists will be able to build a factory to manufacture gold.
>
> No baseball game will be called off on account of rain in the Era of Atomic Energy. No airplane will bypass an airport because of fog. No city will experience a winter traffic jam because of snow. Summer resorts will be able to guarantee the weather and artificial suns will make it as easy to grow corn and potatoes indoors as on a farm. (Ford 1982, 30–31)

It was in this atmosphere that President Harry S Truman in 1946 signed the Atomic Energy Act, creating a new agency, the Atomic Energy Commission (AEC). The AEC would be responsible for both promoting the peaceful use of atomic power and ensuring its safety.

Nuclear Power Comes On-Line

The early development of nuclear power would be controlled primarily by the U.S. government, not by private industry. One reason is that the data on nuclear fission were a closely guarded military secret. In addition, private companies were reluctant to invest in an expensive, untried technology. Thus in 1947 the Atomic Energy Commission established the Reactor Safeguards Committee to develop ways to make nuclear power safe and reliable. The AEC also established an Industrial Advisory Group under chairman James W. Parker to investigate peaceful uses of atomic energy. In the early 1950s, experimental reactors were used to demonstrate the generation of electricity from nuclear power. The United States also offered nuclear technology to its international allies in President Dwight D. Eisenhower's "Atoms for Peace" speech in 1953.

The year 1954 was a watershed year for nuclear power. The first nuclear-powered submarine, *Nautilus,* designed by the energetic Admiral Hyman Rickover, demonstrated the reliability of nuclear energy with a vividness that no AEC paper could match. Its relatively simple reactor system, using ordinary water to both moderate the nuclear reaction and cool the core, would become the standard technology for commercial nuclear power plants. That same year, a revised Atomic Energy Act threw open nuclear technology for private investment and development. Construction of the first commercial nuclear power station began at Shippingport, Pennsylvania. It was in this spirit of overflowing optimism that AEC chairman Lewis Strauss would make a statement that would come back to haunt nuclear power advocates: "It is not too much to expect that our children will enjoy in their homes electrical energy too cheap to meter" (Strauss 1954).

In 1957 the Shippingport plant went into operation, and President Eisenhower signed the Price-Anderson Act. This law assured utilities that were worried about the potentially huge liabilities following a nuclear accident that their costs would be limited. Ever since, nuclear critics point to Price-Anderson as an unjustified subsidy that led to the expansion of an industry that could not have survived in a free market.

By 1960, three commercial nuclear plants were in operation, and by the mid-1960s the nuclear industry really hit its stride, with fifty nuclear plants on order, totaling forty thousand megawatts of generating capacity. By now, nuclear plants had practically become commodities, sold at fixed prices.

Nuclear power seemed to be the wave of the future. Even the Sierra Club, the premier U.S. environmental group, praised nuclear power as a clean energy source and became one of the strongest advocates for building the Diablo Canyon nuclear plant in California. Responding to concerns about fossil fuel pollution, McChesney Martin, chairman of Florida Power and Light (FP&L), said he would never build another coal plant.

Radiation, Health, and the Environment

Despite the great enthusiasm for nuclear energy, the potential danger inherent in the radiation that it depended on could not be escaped. Radiation can make people sick and sometimes kill them, either immediately or years later through cancer. It can also cause genetic mutations leading to crippled offspring. Yet radiation is not really exotic: It's a pervasive part of the natural environment.

Radiation doses are measured in rem ("radiation equivalent man") or millirem (a thousandth of a rem). On the average, people are exposed to about 360 millirem per year from natural background radiation. This radiation comes from cosmic rays penetrating the atmosphere from space as well as from small amounts of radioactive material in the rocks, soil, and building materials that surround us.

Several factors can influence the amount of background radiation a person receives. People living in Colorado, for example, receive about seventy millirem more in background radiation than those living in Florida, because the higher altitude in Colorado means that there is less atmosphere to shield people from cosmic rays. (This also explains why crews on commercial airliners actually have a higher annual radiation exposure than workers in nuclear power plants.) People living in houses built on soil that produces the radioactive gas radon from decaying traces of uranium can also have greater-than-average exposure, for the gas tends to accumulate in basements.

People receive artificial radiation from a number of sources—mainly medical X-rays, but also consumer products such as smoke detectors and clocks with illuminated paint on their dials. Compared to natural sources, a nuclear power plant that operates without accident adds very little to the radiation already in the environment. It exposes people in the surrounding fifty miles to only an additional 1.5 millirem.

Radiation can damage living tissue in several ways. High doses of one hundred rem or more physically disrupt and destroy cells, leading to radiation sickness, with symptoms such as vomiting, intestinal bleeding, and anemia (destruction of red blood cells). Half of all persons exposed to four hundred rem of radiation will die. Smaller doses may have no immediate effect but can lead to an increased risk of leukemia or other forms of cancer. Finally, radiation exposure of the reproductive cells can damage the DNA sequences, leading to mutations whose effect on offspring can range from insignificant to lethal.

Given these potential effects, and given the amount of radiation already present in the environment, how much additional human-made radiation is too much? One theory says that small doses below some as-yet-unknown threshold have no significant health effect. However, some scientists believe that any dose of radiation, even a small fraction added to the background dose, introduces some additional risk. Recent research suggests that this "linear" relationship between dose and effect is more likely to be correct, at least until the dose becomes very small.

The study of "cancer clusters," geographical areas that show an unexpected number of cases of leukemia (cancer of the blood), has further fueled the controversy. In the 1980s cancer clusters were discovered in the vicinity of the Pilgrim nuclear power plant in Massachusetts as well as near several plants in Great Britain. Nuclear critics argue that these clusters arise from the effects of added radiation exposure from the plants, but it has proven difficult to determine whether the clusters are statistically significant and, if so, whether they really correlate to the small amounts of radiation emitted by the plants. (Paradoxically, at Pilgrim, the largest "cluster" turned out to be in the study area farthest from the plant.)

Further, risk cannot be assessed in a vacuum. Life has hundreds of quantifiable risks, and that leads to difficult trade-offs. Physics professor Richard Wolfson points out that "you might consider moving out if a nuclear power plant were built in your neighborhood. On the other hand, suppose your new location requires that you drive 10 miles further in your round-trip commute to work each day. In 10 years of 250 working days each, the risk associated with the extra driving will give an average lifespan reduction of nearly 7 days—more than triple the risk of staying next to the nuclear plant" (Wolfson 1993, 73, 75).

The Search for Safety

Awareness of the danger of radiation exposure led to concern about the safety of the growing number of nuclear power plants. As with many areas of controversy, the early defenders of nuclear power tended to make earnest, reassuring statements. For example, a brochure from the Nuclear Energy Institute explains that

> [the "safety culture"] is a way of doing things that's ingrained in everyone from the plant manager to an apprentice welder. It's what makes a technician go to his supervisor to suggest a safer way to fix a leaking pipe. It's why every mechanic is expected to follow procedures that produce less radioactive waste.

> The safety culture is the foundation of safety at every nuclear plant. The managers and workers take it very seriously. (Nuclear Energy Institute, n.d., 10)

Nuclear critics, however, point to a history of less-than-forthright statements by the nuclear establishment. Back in 1957, at the same time that it was assuring the public of the safety of the new nuclear power plants, the Atomic Energy Commission's Brookhaven National Laboratory issued a secret report that assessed the potentially devastating effects of a nuclear accident that might kill thousands of people, cause $7 billion worth of damage, and poison fifty thousand square miles of land. And in 1977 a researcher at the Union of Concerned Scientists discovered the "nugget file," an internal file kept by an official at the Nuclear Regulatory Commission (NRC) that lists hundreds of instances of potentially serious equipment failures and operator errors.

Nuclear plant safety involves both the reactor core and the surrounding cooling systems. During the 1950s, builders of the early reactors were most concerned with keeping the nuclear reaction under control, poised at the level of neutron production where each fission event led to an average of one additional event, resulting in steady power output. Such control can be difficult because conditions of temperature and pressure and the presence of fission products are constantly changing the absorption of neutrons and thus the rate of fission. However, an emergency system monitors these conditions, and if the reactor is in danger of going "supercritical" a safety device automatically inserts neutron-absorbing control rods into the core, halting the chain reaction. This event is called a "scram."

The importance of the cooling system arises because the radioactive decay of uranium and other radioactive atoms in the core cannot be shut off. Even when a fission reaction is no longer possible, heat will still build up in the core because of this inherent radioactivity. As long as coolant (water, in most reactors) continues to circulate through the core, the heat is carried away and there is no problem. But if the supply of coolant is cut off, such as by a pipe break, a pump failure, or loss of electric power to the pump, some backup cooling or power system must take over, or a "loss of coolant accident" (LOCA) will occur.

In a LOCA, as the core heats, it turns whatever water remains in the core to steam, dissipating it. As the core itself becomes exposed to the air, it heats rapidly and the fuel itself may start to melt. This is the infamous "meltdown" that can potentially result in the molten fuel rupturing the reactor vessel and even the containment building. The worst-case result is a massive release of highly radioactive liquid or steam into the environment, possibly accompanied by an explosion or fire.

In 1963 researchers began to build the LOFT (Loss of Fluid Test Reactor) in order to determine exactly what takes place after coolant is lost in a reactor core. However, the project fell victim to bureaucratic delays and was never used for its intended purpose. Dr. Carroll Wilson, the first general manager of the AEC during the 1950s, later believed that "one of the grievous errors of the Atomic Energy Commission in the 1960s was the failure to carry out the repeatedly recommended experiment of running a [large-scale] reactor to destruction to find out what would really happen instead of depending on studies based on computer models for such vital information" (Norton 1982, 21).

Meanwhile, the government was revamping nuclear regulation. It had been recognized by policymakers that the old Atomic Energy Commission had two conflicting mandates: to promote the widespread use of nuclear power, and to regulate to ensure its safe use. Given the realities of political pressures and bureaucratic life, it seemed inevitable that the agency could not satisfy both the industries it was supposed to regulate and the public interest. In 1974, Congress took a different approach by creating the Nuclear Regulatory Commission, which was intended to focus only on the regulatory task. Its responsibilities include:

- licensing of nuclear reactors (both initial construction and operation)
- regular inspection of reactor facilities
- setting technical specifications for reactor and facility components
- procedures for inspecting components as part of the manufacturing process (quality assurance)
- emergency preparedness for potential accidents or natural disasters affecting nuclear plants
- standards for training and supervision of nuclear plant employees.
- standards for disposal of low-level radioactive waste (incineration) and temporary storage of high-level waste such as spent nuclear fuel, as well as transportation of nuclear materials

- the licensing of uranium enrichment facilities
- decontamination and decommissioning of nuclear facilities

The NRC does strictly regulate and monitor the operations of nuclear power plants, but its oversight has not prevented significant accidents and other troubling events from occurring.

Three Mile Island

Some accidents in experimental power plants had occurred in the 1950s and 1960s, but the 1970s would bring the biggest challenge to the U.S. nuclear industry's promise of safe operation. The first major incident came in March 1975, when an electrician at the Brown's Ferry nuclear power plant in Alabama was using a lighted candle to check for air leaks in the cable room, where there were big bundles of electrical cables supplying power to the plant's many systems. He accidentally started a fire that rapidly burned through some of the cables, cutting off power to the controls for the safety systems for both of the plant's reactors. The reactors automatically scrammed and shut down, but the reactor cores continued to produce heat from residual radioactive decay. The water covering the cores boiled away, and in one reactor the core was within four feet of being uncovered, which could have led to the fuel melting in a disastrous meltdown.

Fortunately, technicians were able to jury-rig a pump and restore a flow of coolant before that happened, and no significant radioactivity was released to the environment. Plant officials said that this showed how the plant's multiple safeguards could prevent disaster even when a major safety system was disabled. But a senior project manager at the AEC and NRC, Robert Pollard, resigned after the fire in protest at the potential dangers not being taken seriously. By the mid-1970s some engineers had also resigned, deciding that they could no longer support nuclear power. But worse was to come.

In the early morning hours of March 28, 1979, a valve in the condenser system in one of the steam generators at the nuclear power station at Three Mile Island, Pennsylvania, failed. This stopped the flow of water through the plant's secondary cooling system. A backup pump automatically started up to keep the flow of water going. Unfortunately, the valves connecting this backup system to the pipes going to the core had been accidentally left closed, blocking the flow of water. To make things worse, someone had left a paper tag on the reactor's control panel that covered up the light that would have told operators about the problem. Without cooling water, the temperature and pressure in the core began to rise rapidly. As designed, in only a few seconds the reactor's pressure relief valve opened to release the excess pressure, and the reactor scrammed to stop the chain reaction. So far, despite an equipment malfunction and a human error, the reactor's multiple safety systems had prevented a serious problem.

Unfortunately, as the pressure in the core fell, the relief valve for the main cooling system failed to close, despite a light in the control room indicating that it had done so. Heat began to build up again as the coolant continued to escape as steam from the reactor vessel into the surrounding containment structure. This triggered the emergency cooling system, but the operators, having no indication that the relief valve was open, assumed that there was now too much water in the reactor. They turned the flow in the emergency cooling system down to a trickle. The operators, confused but not panicking, worked for about two hours to try to reduce the water level, not realizing that the water was coming from the overheated core and they were experiencing a LOCA. When the operators tried to reduce the water buildup by turning off the pumps that circulated water in the reactor core, the remaining water boiled away and exposed the core. The steam reacted with the exposed metal cladding of the fuel rods, producing highly inflammable hydrogen gas. Meanwhile, radioactive fission products were escaping into the containment building, setting off radiation alarms. Finally, the operators realized the relief valve was stuck, and they tried to restore the flow of coolant. But they didn't realize the core had been uncovered.

Confusing bits of information began to leak out into the surrounding community. Local officials began to realize something was wrong at Three Mile Island, but they couldn't get understandable answers from nuclear officials. As water from the reactor overflowed holding tanks in the containment structure and operators released steam to relieve pressure, radiation began to escape into the environment. Suddenly, at 2 P.M. that afternoon, instruments detected a sudden, brief rise in pressure. Unknown to the operators, the accumulating hydrogen in the reactor vessel had created a small explosion.

Later in the day, the operators believed they had regained control of the reactor. By then, however, news of the radiation release had panicked people in the surrounding area. Public alarm only increased when nuclear officials began to worry that a further buildup of hydrogen in the reactor might lead to a major explosion that would spew highly radioactive material for miles around. Plans for evacuating the nearly one million people living within twenty miles of the plant were considered, but the NRC then determined that there was no danger of an explosion.

About ten years later, NRC investigators digging down into the ruined reactor discovered that the top of the core had sunk five feet into the reactor vessel. There had indeed been a meltdown, and one worse than they had anticipated. But the reactor vessel, though cracked, had held despite the heat of twenty tons of melting uranium fuel at a temperature of up to 5,000 °F.

What is the lesson of Three Mile Island? For nuclear critics and much of the public, it seemed clear enough: Nuclear power was dangerous and prone to getting out of control, operators were fallible, and officials were unwilling to give straight answers. Supporters of nuclear power, however, noted that the containment had held. Despite the meltdown, only a relatively small amount of radiation had escaped into the environment.

The massive cleanup effort at Three Mile Island would cost almost $1 billion from utilities, their customers, and taxpayers—and still leave questions about the safety of the site because some radioactive material will be left behind. Also left behind is a roomful of studies that have failed to answer the question of why Three Mile Island hadn't melted down completely or blown up from a hydrogen explosion. Probably the most significant fallout from the near-disaster, however, was the effect on public confidence in the U.S. nuclear industry.

The Chernobyl Disaster

Public opposition to the building of new nuclear power plants combined with the lower prices for alternative fuels such as oil and coal had already slowed the expansion of the nuclear power. However, other nations—notably France and Japan, which have to import their oil—had embraced the atom as an energy alternative. The Soviet Union, too, had built an extensive network of nuclear power plants and was also selling them to countries in Eastern Europe. But events would soon reveal fatal flaws in the Soviet nuclear industry.

History's worst nuclear accident began on the night of April 25, 1986, at the Soviet nuclear power plant at the village of Chernobyl. The operators were conducting an experiment with the steam turbine to see whether in the event of a power failure the blades could still "coast" on inertia long enough to drive an emergency cooling pump. The experiments were running behind schedule, so the operators were rushed. They were too quick in throttling power output from the normal service level down to the 25-percent level needed for the experiment. Shutting down a reactor too fast creates products that absorb neutrons and "poison" the core so that it can't be restarted for many hours. As the reactor became poisoned, operators tried to compensate by pulling the control rods nearly completely out to speed up the chain reaction. Even so, they achieved only 6-percent power. Operations under these conditions are very unstable and indeed were forbidden by the operating rules, but the plant's supervisors and operators did not seem to take regulations seriously.

Despite the danger, the supervisors ordered the experiment to continue. At 1:19 A.M. they turned on the water pumps that were to be tested. Because water absorbs neutrons, it made the "poison" situation worse, and the operators pulled out the rods even further to keep the reaction going. The reactor was now in a very dangerous condition where, if there was a loss of water, the neutrons from the now uninhibited fission could send the reaction out of control, doubling the power every second or so.

At 1:21 A.M. the added water flow was stopped, but operators failed to immediately reinsert the control rods. The computer issued a warning that the reactor was now unsafe and should be shut down, but for some reason the operators ignored the message. Meanwhile, the pumps, being driven only by inertia, slowed down and reduced the water flow still further. Now things really got out of control: Less water meant more heat, more heat meant more steam, more steam meant less water—and fewer neutrons being absorbed, leaving more to accelerate the nuclear chain reaction.

Only about three minutes later, operators began to frantically reinsert the control rods, but that system operated more slowly than those in American reactors, and things were getting worse every fraction of a second. Damage to the core from heat began to block the rods so they couldn't be inserted all the way.

Soon the heat level was one hundred times the reactor's design level. At 1:24 A.M. two loud explosions occurred, and a glowing mass flew out of the top of the reactor building. The reactor had blown up not like a nuclear bomb but like a giant overheated boiler—a boiler filled with deadly radioactive materials. The explosions were followed by fires. Firefighters heroically put out the fires, many receiving lethal radiation exposures in the process. Helicopters dropped neutron-absorbing boron, clay, lead, sand, and dolomite over the top of the building to dampen the radiation—some of the pilots, too, would die from radiation sickness.

Despite the beginnings of the era of *glasnost,* or "openness," Soviet authorities reacted in the old Soviet way: secrecy. The first warning of the Chernobyl disaster came when radiation monitors in Sweden, Finland, and other neighboring countries signaled an abnormal level of radiation in the air. Eventually the Soviet authorities revealed what had happened, and international teams offered assistance in treating the victims of radiation. People in the surrounding area were evacuated, but they have a somewhat higher risk of cancer now and in the future.

The extent of the disaster shocked people around the world and helped solidify opposition to nuclear power, which combined with the anxiety about possible nuclear war that also seemed to be peaking at the time. Nuclear experts in the United States, however, pointed out that the Chernobyl reactor was different from U.S. power reactors in several key respects. Instead of using water as both moderator and coolant, it used graphite as a moderator. With water, a loss of flow means a loss of moderator that automatically slows down the nuclear reaction by reducing the number of available neutrons. With a graphite moderator a loss of coolant water doesn't reduce the flow of neutrons, and indeed, the graphite is flammable and likely to add to the problems by catching on fire. Further, the Soviet reactors had only an ordinary building enclosing them, not a thick, reinforced-concrete containment structure as in U.S. power reactors. Finally, U.S. experts noted that the lax behavior of workers and the ignoring of regulations typical at Chernobyl were not typical of U.S. nuclear plants, where any unusual experiments could be carried on only after lengthy planning and under close supervision.

In the 1990s the legacy of Chernobyl survived the fall of the Soviet Union. Fifteen other reactors in Russia and the former Soviet countries use the same RBMK graphite-moderated design as the Chernobyl plant. Already suffering from severe economic strain, Russia finds it very difficult to phase out these plants, and though a number of European nations and the United States have offered technical and financial assistance, there has been no comprehensive plan to deal with nuclear safety. One proposal is to replace the nuclear plants with natural gas plants, but this would cost at least $18 billion plus the cost of the gas itself.

Reacting to the impact of Three Mile Island and Chernobyl, reactor designers have sought to reach a new level of safety. Although more backup systems such as pumps and generators can be added, such precautions increase both the cost and the complexity of the system. Instead, designers are putting reactors on the drawing board that use the laws of physics to build in additional stability. For example, by building a lower-power (600 megawatts instead of 1,200 megawatts) reactor, the heat is reduced to the point where it can be handled by a simple system where water flows in through gravity and circulates by convection (warm water rising because it is lighter). The nice thing about gravity is that you don't have to worry about it failing. The fuel core itself can also be redesigned so the fuel pellets are made smaller and are encapsulated by ceramic materials that can withstand extremely high temperatures. The fuel would be arranged in such a way that it would not melt even after a total loss of coolant. Critics, however, have argued that these new designs may not be as safe as claimed, and that adopting them might create a level of overconfidence that would lead to a slackening of regulation, oversight, and monitoring.

The Nuclear Waste Problem

Even if a nuclear plant operates perfectly with no escape of radiation to the environment, there is still the problem of what to do with radioactive waste. Understanding the problem of nuclear waste begins with a look at how nuclear fuel is made and used, and what happens to it after it is no longer usable in the reactor.

First, uranium must be mined. (In the United States, uranium is mined mainly in the western and southwestern states.) The uranium is obtained either by the ore being dug out and milled (crushed and ground) or by the uranium being leached, or dissolved within the rock and percolated up and collected. This latter method reduces environmental impact by minimizing the accumulation of tailings (leftover ore that contains some radioactive uranium). The uranium is shipped from the mine in the form of a powder called "yellowcake." (A ton of ore yields about four pounds of yellowcake.)

The yellowcake goes to a conversion plant where it is chemically changed into uranium hexafluoride gas, which is shipped in cylinders to an enrichment plant. (The three enrichment plants in the United States are in Oak Ridge, Tennessee; Paducah, Kentucky; and Portsmouth, Ohio.) The enrichment plant uses a process called gas diffusion to separate out the lighter, faster-moving particles of the fissionable uranium-235 from the heavier, nonfissionable uranium-238. This raises the concentration of uranium-235 from its natural level of less than 1 percent to 3.5 percent, which is enough to sustain the fission chain reaction in a nuclear power plant.

The enriched uranium is then fabricated into fuel by being compressed into pellets about the size of a fingernail. The pellets are inserted into zirconium tubes (rods) about twelve feet long. The rods are capped, sealed, and stacked to form a reactor core consisting of about fifty thousand rods and weighing about one hundred tons.

As the fuel undergoes nuclear reaction to generate power it gradually accumulates fission products—the highly radioactive isotopes that result from the uranium atoms being split. Eventually the presence of these products reduces the efficiency of the nuclear reaction to the point where it can no longer generate adequate power. The material in the core is now considered to be "spent fuel." It contains radioisotopes such as cesium-137 and strontium-90 that pose great hazard if they get into the environment, where they can be absorbed by the bodies of humans or animals.

It is possible to "reprocess" spent fuel and extract the unexpended uranium-235 as well as the plutonium-239 that had been created by the chain reaction. (After a year's operation a typical reactor core will have about two hundred pounds of uranium and five hundred pounds of plutonium.) The process involves a tedious, painstaking chemical separation of these fissionable elements from the waste. Once recovered, the uranium could be used to make more reactor fuel, and the plutonium could be used in a "breeder reactor" also to make more fuel. Despite the advantage of being able to eke out the supply of uranium, reprocessing has not been adopted in the United States partly because of concerns that the pure uranium and plutonium would be attractive to terrorists.

As a result of lack of agreement on how to proceed, much of the spent fuel in the United States has ended up being stored at the reactor site, immersed in a steel-lined pool where the water provides both cooling and radiation shielding. For every one thousand megawatts of annual power production, a nuclear plant produces about twenty metric tons of spent fuel. By 1995, about twenty-eight thousand tons of spent fuel were stored in the United States; that number is expected to increase to forty-eight thousand tons by 2003. This spent fuel is also known as high-level radioactive waste. It amounts to only about 0.1 percent of radioactive waste shipments, but it represents the greatest risk in case of accident.

A number of rather fanciful proposals have been made for getting rid of nuclear waste for good. One is to shoot it into space, where it would escape the pull of earth's gravity and fall into the sun, certainly the ultimate in incinerators. But the failure of six Russian nuclear-powered space probes (including one that carried half a pound of plutonium and crashed into the ocean off the coast of Chile) has underlined the dangers of this approach, not to mention the very high cost of sending anything into space. Other proposals include burying the waste beneath the ocean floor or sinking it beneath Greenland's permanent ice sheets. But the only really practical solution is to find a more down-to-earth place where the waste can be stored without leaking or coming into contact with water.

The Nuclear Energy Institute has estimated that twenty-five U.S. nuclear plants would run out of on-site storage space by the year 2000. Congress required that a temporary waste facility be ready by 1998, but in the Nuclear Waste Policy Act of 1982, the Department of Energy agreed to accept shipments of spent fuel from commercial nuclear reactors as of January 31, 1998.

Since the act was passed, utilities have charged their customers a penny for every kilowatt-hour generated by a nuclear reactor, creating a fund to be used for the development of a permanent waste disposal site. This fund has now grown to more than $10 billion. But finding a place to put the waste sites has become a political nightmare.

Yucca Mountain

After a number of proposals fell through, in 1987 Congress decided to focus all its waste disposal efforts on one site, Yucca

Mountain, Nevada, about one hundred miles northwest of Las Vegas. This huge site will have more than a hundred miles of underground tunnels about one thousand feet below the surface, and eight hundred feet or more above the water table. But scientists must conduct a long, painstaking testing program to try to prove that the stored waste cannot leak into the surrounding rock and potentially reach the water table, where it could get into the area's water sources. Some scientists, such as two physicists at Los Alamos and a group at the Energy Department's Savannah River research facility, believe it is possible that thousands of years from now, when the metal waste containers have crumbled, the still very radioactive waste could pile together with rocks that will moderate the neutrons from the radioactive atoms, causing an explosive chain reaction. (Scientists have discovered that a rock formation in Africa functioned as a natural nuclear reactor in prehistoric times.) However, other scientists, including a review panel at the Los Alamos National Laboratory, say that the chance of this happening is remote, and the chance of a chain reaction within a fuel canister is also very low (Yucca Mountain Project, 1999).

By 1995, when it seemed clear facilities would not be ready in time, nineteen utilities and forty states and regulatory agencies filed suit, claiming that the federal government had reneged on its agreement after the utilities had paid millions of dollars. As of April 2000, the Yucca Mountain site remains in limbo, with President Clinton vetoing legislation that would have allowed waste storage to begin.

The delays in operating the Yucca Mountain site have led some members of Congress to push for the building of a large aboveground temporary storage facility. One possibility arose when the Mescalero Apache tribe in New Mexico agreed, after a contentious debate and two rounds of voting, to accept a "monitored retrievable storage" facility on its land in exchange for payments from the thirty-one utilities that have expressed an interest in using the facility.

Transporting Nuclear Waste

Once a disposal site is established, the waste has to be shipped there safely. Each year about three million shipments of radioactive materials (mostly low-level) are made throughout the United States. The nuclear industry points out that the safety record for these shipments is better than that for other forms of hazardous materials and that releases of radioactive materials to the environment are rare.

A typical shipment cask of high-level waste consists of spent fuel rods sealed in a stainless-steel cylinder and encased in heavy metal shielding and additional layers of steel. These casks are large (about five feet in diameter and seventeen feet long) and weigh up to forty tons (if shipped by truck) or as much as 70–100 tons (if shipped by train). The NRC subjects these casks to brutal tests, including an 80-mph crash into a thick concrete wall, a collision with an 80-mph locomotive, drops of up to two thousand feet onto hard ground, and burning at 2,000 °F for more than an hour and a half. The actual shipping is regulated by the Department of Transportation (DOT) and regulators in the states through which the shipment will pass.

Another kind of nuclear waste is much less radioactive than spent fuel, but also much bulkier. It is called low-level waste. Low-level wastes are materials that are weakly radioactive or have been contaminated with relatively small amounts of radioactive material. Radioactive materials are used in a variety of applications that lead to the generation of a stream of low-level nuclear waste. These include radioisotopes used in hospitals for diagnostic tests and cancer treatment as well as radioactive tracers used to test materials in factories or to analyze substances in laboratories. Some consumer products, such as illuminated watch dials and smoke detectors, contain small amounts of radioactive material. Other low-level waste includes gloves and protective clothing worn by medical or nuclear power plant workers. About two-thirds of low-level waste comes from the nuclear industry, including uranium mining and refining, fuel fabrication, and the operation of nuclear power plants. Use of techniques that minimize waste creation has reduced the amount of low-level waste by about half since 1980, but it still amounts to millions of cubic feet annually.

The Low Level Waste Policy Act of 1980 made each state responsible for the disposal of wastes generated within the state. Because one facility could handle wastes from several states, there has been an attempt to build such regional facilities. However, the familiar response of NIMBY ("not in my backyard") means that the neighbors of any such proposed waste site will generally oppose it vociferously.

For example, in 1993 California approved a low-level waste storage site at Ward Valley. But because that site was on federal land, Interior Secretary Bruce Babbitt was able to insist on an independent review by the National Research Council, together with the Bureau of Land Management and the U.S. Geological Survey. But this complex environmental review and regulatory process makes it possible to bog down a proposed site almost indefinitely. By the late 1980s only three low-level waste sites were in operation, in Barnwell, South Carolina; Beatty, Nevada; and Richland, Washington.

Nuclear Proliferation

The fuel used in nuclear reactors (about a 3.5-percent concentration of uranium-235) is far too dilute to be used to make a nuclear bomb, which requires enrichment to a level of around 30 percent or higher. But it is certainly possible that a "rogue" nation or even a well-financed terrorist group could, if it obtained commercial nuclear fuel, enrich the fuel using the same processes discovered in the 1940s during the Manhattan Project.

Further, high-level waste (spent fuel) might be clandestinely reprocessed to extract the uranium-235 or plutonium-239, either of which could be used to make a nuclear weapon. To reduce the proliferation of nuclear weapons to nonnuclear nations or terrorists, the United States has agreed to accept spent fuel from some forty nations with nuclear research reactors, including Canada, Japan, and Australia.

With the end of the Cold War, the world is suffering a plutonium glut. The dismantling of former U.S. and Soviet nuclear weapons is likely to yield one hundred tons of plutonium, and the recovery of plutonium from spent reactor fuel amounts to about twenty tons per year. In highly concentrated form, plutonium is compact and relatively easy to conceal and transport. This means that a terrorist group might obtain the few kilograms needed to create a "low-yield" but still devastating nuclear weapon, using plans that are now widely available.

One way to forestall this threat is to mix the plutonium with uranium into a "mixed oxide" fuel that could not be immediately used for nuclear weapons but could be used to generate electricity in standard nuclear reactors. Unfortunately, the cost of producing this fuel is currently about seven times that of conventional uranium fuel. Alternatively, the plutonium could be irradiated in a reactor and combined with fission products so that it would be very difficult for someone to process it for weapons use. Finally, plutonium that has been extracted from power reactor fuel could be "un-reprocessed" by mixing it back into the spent fuel wastes from which it was extracted.

Another proposal would use surplus military plutonium in a "triple play" reactor that would simultaneously deplete the plutonium, produce tritium (a radioactive gas needed for new nuclear warheads), and generate electricity. But this proposal is opposed by opponents of both nuclear power and the further development of nuclear weapons.

The United States and Russia face years of tricky negotiations in trying to come up with a plan that prevents diversion of plutonium while storing it safely or finding practical uses for it.

What Do You Do with an Old Nuke?

Although cleaning up in the aftermath of a Three Mile Island or a Chernobyl presents special difficulties, every nuclear power plant will reach the end of its useful life at some point. The NRC allows utilities to choose one of three alternatives for decommissioning an unwanted nuclear plant.

The "Decon" option involves hauling away the reactor vessel and radioactive components and decontaminating the area so that it can be used for any normal purpose. Because most of the power-generating machinery at a nuclear plant is not radioactive, it is possible to remove the reactor vessel and replace it with a conventional coal or gas boiler while reusing the turbines for power generation.

The choice to "Entomb" means to seal the reactor in concrete or other shielding material—this has been done with some government reactors. Finally, the "Safstor" option can be chosen: The reactor can be mothballed, allowing it to run down gradually under continuous supervision over a period of years until it has lost most of its radioactivity.

By the mid-1990s twenty-one U.S. nuclear plants had been shut down after an average of only eleven years of operation, either because they were not economical or because fixing problems would not be cost effective. About a quarter of the remaining reactors may also be shut down. So far, utilities have chosen the Safstor mode, despite its high annual costs, because there is currently no place to store the wastes that would be removed during full decontamination.

Reactors in the United States have been licensed for forty years. If they reach the end of their license period, the last reactor built in the United States (Watts Bar 1 in Tennessee) would run until 2036. But most reactors are likely to be shut down well before then. In fact, only one utility company, Baltimore Gas and Electric, which operates the Calvert Cliffs plant in Maryland, has actually applied to renew its license beyond forty years. Such license renewals are expensive, taking years in applications and modifications and costing more than $10 million, with no guarantee that the renewal will be granted.

Given the increased competition between utilities in the era of deregulation, however, keeping a well-performing nuclear power plant running may be a good decision. According to Dan Ford, certified financial analyst with HSBC Securities, Inc., "the utility can continue using an approved, fixed power asset to generate revenue and has the financial option of stretching out the time it takes to repay the plant's initial investment. Furthermore, the costly chore of decommissioning an environmentally sensitive facility can be postponed" (Price 1999, 17).

Sometimes reactors have been sold to another company that can continue to operate them economically rather than decommissioning them. But Jim Riccio, attorney and researcher for Ralph Nader's Public Citizen group, says that the selling of nuclear plants like Three Mile Island unit 2 and Pilgrim for a tiny fraction of their original cost is no bargain for consumers or taxpayers:

> This is what Ralph [Nader] calls "lemon capitalism"—when an investment risk turns bad and the public is maneuvered into picking up the tab. Many of these reactors met with massive public resistance. Making the public pay for them now is just sickening. There are still over 100 nuclear power plants in the country that will probably be working for 10 or more years. We're entering into a more dangerous period in nukes. These are aging reactors. It's a dying industry, so you're not getting the best and the brightest in this field anymore. A lot of the expertise of the previous generation is retiring. The guys who built these plants are retiring. And these are very old utilities that are being placed in

an environment of increasing pressure to compete. (Clarke 1999)

In other words, competition, by leading to pressure to cut costs, may endanger safety at a time when the infrastructure is getting older and less reliable.

A Nuclear Sunset?

With no nuclear plants ordered since 1974 in the United States, it is clear that growth in the American nuclear industry has stalled if not reversed. One reason is obviously the decline in public confidence in nuclear energy following the Three Mile Island and Chernobyl accidents. Psychology professor Paul Slovik asked four groups of people to rank the riskiness of a list of activities. The group composed of experts placed nuclear power twentieth on the list, below things such as surgery and flying in small airplanes. A group of business executives raised nuclear power to eighth on the list. Two other groups, members of the League of Women Voters and college students, ranked nuclear power as the riskiest activity. And according to a March 1999 Associated Press poll, public support for nuclear power is down to 45 percent from 55 percent ten years earlier.

As the antinuclear reaction fueled by Three Mile Island continued, a growing number of nuclear plants that had been in the news began to shut down, including San Onofre unit 1, Rancho Seco, and Humboldt Bay in California; Trojan in Oregon; Yankee Rowe in Massachusetts; and Shippingport, Pennsylvania, which had been the very first one built. The Shoreham nuclear power station in Long Island had just completed low-power testing when opposition by local and state officials forced it to shut down.

But rising costs and the changing economics of energy in the late 1970s and 1980s have also played an important part in nuclear energy's apparently dismal future. Several factors led to nuclear power costs increasing rapidly in the 1970s and early 1980s. The nuclear industry stopped receiving government subsidies (except with regard to liability), and new regulations imposed significant costs.

The deregulation of the power industry in general (removing subsidies and bringing different kinds of power into direct competition) has made nuclear power less attractive. Former Energy Secretary Hazel O'Leary notes that "when one looks from the perspective of a business to open competition, when you look at cutting costs—to ask what are your biggest cost liabilities in the U.S. power industry it is, by and large, nuclear power plants" (Edwards 1999, 68).

As nuclear plants age, equipment failure begins to take an expensive toll. In 1986 a corroded pipe carrying superheated water in one reactor burst open and scalded four workers to death. The NRC imposed new regulations requiring close monitoring of all pipes and replacement of those showing corrosion or brittleness. The heat exchangers, which transfer heat from the core of the reactor to generate steam for the turbine, are also prone to cracking and leaking, sometimes releasing radioactive steam or water into the environment. The leaks are detected by monitoring systems, minimizing such releases, but following a release, the plant must be shut down and expensive replacements made. (Baltimore Gas and Electric plans to spend $300 million to replace its steam generator as part of its license renewal, whereas Portland General Electric decided to close its Trojan plant rather than pay for expensive replacements.) Even the reactor core itself becomes brittle after years of neutron bombardment, reducing its ability to contain melting fuel after a loss of coolant accident.

As a result of these growing expenses, the power that was to have been "too cheap to meter" has sometimes become too expensive to supply. At the same time, following the end of the oil crisis of 1973, oil prices have remained relatively low, and despite pollution control costs, coal, too, has remained a viable alternative. According to energy analyst Barry Abramson of Prudential Securities Research, Inc., "nuclear plants sort of ground to a halt years ago. The fact is that there are low-risk alternatives" (Field 1995, 31).

Since 1974, U.S. utilities not only did not order new nuclear reactors, they canceled the construction of 121 facilities they had previously ordered, having spent $50 billion (in 1995 dollars) on them. Many of these costs were passed on to consumers in the form of higher electricity bills.

In Europe, however, where fossil fuels must be imported in greater proportion, nuclear power continues to be more attractive. France gets about 70 percent of its electricity from nuclear power, and Belgium gets about 60 percent. But in what may prove to be a serious blow to the international nuclear industry, in June 2000 Germany announced that all of its nuclear plants would be phased out over a twenty-year period. Elsewhere, the picture is mixed: Italy has already closed its three nuclear plants. Finland is considering building a fifth reactor, and Sweden is modernizing its existing facilities.

South Korea, Taiwan, and Japan continue to expand their use of nuclear power, using mostly U.S.-designed reactors. Japan, the only nation to suffer atomic bomb attack, now gets about 35 percent of its power from nuclear reactors and is designing breeder reactors and fuel-reprocessing facilities.

A Nuclear Renaissance?

There is another side to the nuclear picture. At the time of the oil embargo in 1973, about 17 percent of U.S. electrical energy was produced from oil and only 5 percent from nuclear energy. By 1993, however, oil was producing only 3 percent, and nuclear energy had risen to 19 percent. By reducing demand for oil for electric production, nuclear power had helped reduce America's dependence on foreign oil supplies.

A report by the DOE's Energy Information Administration (EIA) estimates that despite significant increases in energy efficiency in American businesses and homes, U.S. power consumption will increase by 20 percent through 2020. Nuclear power production will increase until about 2006 as plants built during the 1970s are used to full capacity, but by 2010 the decommissioning of many nuclear plants without any replacements being built will lead to gas-fired power replacing nuclear power in second place behind coal. The portion of power generated by nuclear plants is likely to fall to about 8 percent by 2020. Meanwhile, the DOE has estimated that the United States will need to build about 250 new power plants between 1995 and 2010 to keep up with demand for electricity.

A 1,000-megawatt coal-burning power plant burns more than two million tons of coal in a year. Thousands of people die each year from the effects of the pollution from conventional power sources. Ironically, because coal contains radioactive trace elements, even when a coal plant uses scrubbers or precipitators to filter out 95 percent of its particulate emissions, it still introduces more radioactive material into the atmosphere than does a nuclear plant. By substituting for fossil fuel plants, U.S. nuclear plants in 1991 saved 145 million tons of coal, 265,000 barrels of oil, and 1.7 trillion cubic feet of natural gas. Annually, they keep about 420 million tons of carbon dioxide out of the atmosphere.

The growing concern about global warming probably offers the best hope for advocates of a new expansion of nuclear power in the twenty-first century. Emissions from burning fossil fuels (especially carbon dioxide) trap heat in the earth's atmosphere and seem to be leading to a rise in the average world temperature that could have a devastating effect on coastal communities, agriculture, and other areas. In response to this concern, the environmental summit in Kyoto, Japan, put forth a goal of reducing greenhouse emissions to below 1990 levels early in the twenty-first century. But how can this be done while still providing the energy needs of both the developed world and the emerging economies of nations such as China and India?

Of the nonfossil fuel power sources, water power, once considered an attractive alternative, seems to have little potential for further growth. Hydroelectric dams have already been built in the most likely places, and new ones are opposed because of their disruption of the environment. Solar power offers clean energy and useful heating, but it lacks the concentration to feed the needs of cities and large factories. Nuclear fusion power (created by fusing, or combining, light atoms) would release far less radiation than nuclear fission, but after decades of research it has yet to become commercially viable.

Only nuclear fission power seems to offer the combination of concentrated energy and a lack of conventional pollution and greenhouse gases. Stuart Eizenstat, U.S. undersecretary of state, has stated his belief that "nuclear has to be a significant part of our energy future and a large part of the western world, if we're going to meet these [Kyoto emission reduction] targets" (Schimmoler 1999, 8).

Conclusion: An Uncertain Future

In 1995 the Department of Energy noted that "as we approach the 21st century, the future of commercial nuclear power remains uncertain. What was expected to be a cheap source of electricity nearly 29 years ago has become more costly" (Field 1995, 31). Although the energy and environmental needs of the twenty-first century may offer a new incentive to build nuclear plants, any such renewal faces formidable obstacles.

Senator Pete Domenici (R–New Mexico) insists that "we aren't wisely using nuclear technologies. The current anxiety-laden, fragmented state of nuclear policy debate in the country has created this situation. Irrational fears of perceived risks of nuclear technologies prevent us from actions to address real risks and optimize their use" (Marsh 1998, 47). But the history of secrecy, overblown promises, and lack of forthrightness that critics say has characterized the nuclear industry has also played a part in the predicament it now faces.

This book presents a wide array of sources and resources dealing with the history, development, and controversies surrounding nuclear power. Although much of the body of nuclear literature dates back to the intense interest in (and opposition to) nuclear power in the 1970s and 1980s, much new material is emerging in print and on the Internet.

Government planners and policymakers as well as industry officials will continue to face the issues of aging infrastructure, waste, and the emerging problems of nuclear proliferation. The possible scenarios for nuclear power remain uncertain in a new century that faces both tremendous energy demands and critical environmental challenges. It is thus a good time for those of us for whom nuclear power has faded into the background of the daily news to take a fresh look at this complex and vital field.

References

Andrade, E. N. da. 1964. *Rutherford and the Nature of the Atom.* New York: Doubleday.

Clarke, Kevin. 1999. "Three Mile Islands to Go before We Sleep." *U.S. Catholic* (March).

Edwards, Brian. 1999. "The Lights Dim on U.S. Nuclear Power." *The World and I* (August 1).

Field, David. 1995. "U.S. Nixes Nukes, but Asia and Europe Remain Radioactive." *Insight on the News* (May 1).

Ford, Daniel. 1982. *The Cult of the Atom: The Secret Papers of the Atomic Energy Commission.* New York: Simon and Schuster.

Grady, Sean M. 1992. *The Importance of Marie Curie.* San Diego: Lucent Books.

Inam ur Rehman. 1993. "Historical Evolution of Nuclear Technology." *Economic Review* (June).

Marsh, Gerald. 1998. "Nuclear Power, Yes: Nuclear Power Is the Most Environmentally Benign Source of Electricity." *Bulletin of the Atomic Scientists* (March-April).

Norton, Boyd. 1982. "The Early Years," in Michio Kaku and Jennifer Trainer, eds., *Nuclear Power: Both Sides.* New York: Norton.

Nuclear Energy Institute. *Nuclear Energy: Power for People.* Washington, D.C., n.d.

Price, Stuart V. "Is Time Running out for Nuclear Plants?" *Energy User News* (March).

Rhodes, Richard. 1986. *The Making of the Atomic Bomb.* New York: Simon and Schuster.

Schimmoller, Brian K. 1999. "Nuclear Crossroads?" *Power Engineering* (January 1).

Strauss, Lewis L. 1954. "Remarks Prepared for Delivery at the Founders' Day Dinner, National Association of Science Writers" (September 16).

Wolfson, Richard. 1993. *Nuclear Choices: A Citizen's Guide to Nuclear Technology.* Cambridge, Mass.: MIT Press.

Yucca Mountain Project. "Total System Performance Assessment Peer Review Panel." Final Report, February 11, 1999. Available online at http://domino.ymp.gov/va/support/tspa_peer.nsf.

2

Chronology

Following is a chronology of significant events in the development of nuclear power, including technical advances, accidents, and legislation. The chronology runs from the late nineteenth century to the present, with an emphasis on developments in the 1970s and later. Researchers seeking further historical information should consult some of the historical or archival websites described in Chapter 7.

1895 Wilhelm Roentgen, a German physicist, discovers a new form of short-wavelength radiation that can penetrate objects. He calls them X-rays and receives the Nobel Prize in 1901 for his discovery. Besides being very useful to doctors, X-rays will provide a tool that physicists can use to explore the structure of atoms.

1896 French physicist Antoine Henry Becquerel tries to discover whether uranium produces X-rays when exposed to sunlight. After several foggy days, he develops the film anyway and is surprised to find that the uranium produces radiation by itself without any need of sunlight. This discovery of naturally radioactive materials is the first hint of the power locked inside the atom.

1897 British physicist J. J. Thompson discovers the electron, one of three basic particles that make up the atom.

1898 Pierre and Marie Sklodowska Curie, an intrepid husband-and-wife science team, sift through uranium ore, treat it chemically, and discover two radioactive

1898 *(cont.)* elements, polonium and radium. Radium, a powerful radiation source, is soon used for both medical treatments and research into atomic physics.

1900 French physicist Paul Villard discovers gamma rays, an energetic, deeply penetrating form of radiation.

1903 The Nobel Prize in physics is shared by Becquerel and the Curies.

1905 Albert Einstein introduces his theory of relativity. One consequence of Einstein's work is the equivalence of matter and energy, shown in the famous equation $E=MC^2$. This equation shows why the breakdown of small amounts of matter in the atom can lead to a large release of energy.

1908 New Zealand–born British physicist Ernest Rutherford and his assistant Frederick Soddy receive the Nobel Prize for their explanation of alpha and beta radiation and their demonstration that radioactive emission can change one kind of atom into another (transmutation).

1913 H. G. Wells's novel *The World Set Free* foresees a future where atomic energy is used for both peaceful energy and devastating weapons.

1932 James Chadwick, a former student of Rutherford's, discovers the neutron, an uncharged particle in the nucleus of the atom. He receives the Nobel Prize in 1935.

1934 Italian physicist Enrico Fermi bombards uranium atoms with neutrons. Some of the atoms absorb the neutrons and become neptunium, the first artificial element.

1938 German chemists Otto Hahn and Fritz Strassmann bombard uranium with neutrons and are surprised to find they have produced barium and krypton, two much lighter elements. Austrian physicist Lise Meitner and her nephew Otto Frisch discuss these results and conclude that the uranium atom had been actually split in two by the neutron. They apply the biological term "fission" to this process.

1939 Enrico Fermi carries the news of uranium fission to Washington, D.C., and predicts the possibility of a chain reaction. Albert Einstein, responding to concern within the scientific community, writes a letter to President Franklin D. Roosevelt, telling him of the possibility of building a weapon of unprecedented power—an atomic bomb.

1940 Researchers at the Radiation Laboratory at the University of California, Berkeley, create plutonium, an artificial element that can be used as a nuclear fuel or to make nuclear weapons.

1941 President Roosevelt gives approval for a secret scientific team headed by Enrico Fermi, J. Robert Oppenheimer, and Ernest O. Lawrence to begin construction of an atomic bomb. The massive project is given the code name Manhattan District Project.

1942 Researchers build the world's first nuclear reactor under the football stands at the University of Chicago.

1943 The Los Alamos Scientific Laboratory is built in New Mexico, and scientists and technicians begin the design and assembly of the first atomic bomb. The government also acquires land in the Colorado Plateau and begins to contract out for uranium mining operations.

1945 The atomic bomb is successfully tested at Alamogordo, New Mexico.

The United States drops a uranium-fueled atomic bomb on Hiroshima, Japan, with devastating effects. A second atomic bomb, fueled by plutonium, is dropped on the city of Nagasaki. Japan surrenders unconditionally, ending World War II.

1946 President Harry S Truman signs the Atomic Energy Act of 1946, establishing the Atomic Energy Commission (AEC). One of the responsibilities of this body is to promote the peaceful use of atomic energy, such as the design and construction of nuclear power plants. The Joint Committee on Atomic Energy (JCAE) is also established.

1947 The Atomic Energy Commission establishes the Reactor Safeguards Committee. The AEC also establishes an Industrial Advisory Group under Chairman James W. Parker to investigate peaceful uses of atomic energy.

1949 The Atomic Energy Commission chooses a site in Idaho to construct the National Reactor Testing Station.

1951 The Experimental Breeder Reactor 1 in Arco, Idaho, generates the first electricity from nuclear energy.

1952 The National Research Experimental Reactor at Chalk River, Canada, gets out of control, leading to a partial core meltdown that is contained.

1953 President Dwight D. Eisenhower delivers what becomes known as his "Atoms for Peace" speech before the delegates of the United Nations in New York. He calls for an international agreement to share in the development of nuclear technology.

1954 *Nautilus,* the first nuclear-powered submarine, is launched. Its light water reactor, built under the leadership of Admiral Hyman G. Rickover, demonstrates the feasibility of nuclear power and will become the basic design used by most nuclear power plants.

The Atomic Energy Act of 1954 is passed, revising the 1946 act. The new legislation allows for private ownership of nuclear power facilities and encourages the participation of private industry in the general development and application of nuclear energy. The JCAE approves a five-year program for reactor development, and construction of the first commercial nuclear power plant is begun at Shippingport, near Pittsburgh, Pennsylvania.

1955 The Atomic Energy Commission begins the Cooperative Power Demonstration Program, which offers technical assistance and a reduction of some regulatory fees to any utility that desires to build a nuclear power plant.

The Experimental Breeder Reactor begins operation in Idaho, demonstrating the ability to create not only nuclear power but also its own fuel.

The first United Nations International Conference on the Peaceful Uses of Atomic Energy is held at Geneva, Switzerland.

1956 President Eisenhower directs the AEC to make available to other qualifying nations, for sale or lease, twenty thousand kilograms of uranium-235 for use in power and research reactors, and another twenty thousand kilograms for power reactors in the United States.

1957 The Brookhaven Report, "Theoretical Possibilities and Consequences of Major Accidents in Large Nuclear Plants," is released. It seeks to reassure the public and industry officials that most anxiety about nuclear power is unfounded. At the same time, however, a secret report by the same organization estimates that a nuclear meltdown could cause $7 billion in damage, kill several thousand people, and contaminate fifty thousand square miles of land.

President Eisenhower signs the Price-Anderson Act, which encourages utilities to invest in nuclear power plants. The law limits the financial liability of plant owners and contractors in the event of a major accident.

The United Nations establishes the International Atomic Energy Agency (IAEA) to promote the peaceful uses of nuclear energy throughout the world.

The first commercial nuclear power station, at Shippingport, Pennsylvania, begins operation. This demonstration sixty-megawatt plant uses a pressurized-water reactor and is soon supplying electric power to the surrounding community.

1958 Commonwealth Edison begins construction of its Dresden nuclear power station, unit 1, outside of Chicago.

1960 The Atomic Energy Commission releases a new ten-year plan for nuclear power development.

The first boiling-water reactor, manufactured by General Electric, goes into service with the opening of Commonwealth Edison's Dresden Plant 1.

1960 *(cont.)* Yankee Nuclear Power Station in Massachusetts becomes the third commercial nuclear power plant to go into service in the United States.

1961 A test reactor in Idaho Falls, Idaho, goes out of control, resulting in the death of three maintenance workers.

1962 President John F. Kennedy considers a report from the AEC that looks at the nation's growing energy needs and recommends an expansion of nuclear power capacity, including the development of breeder reactors that can create their own fuel.

1963 Jersey Central Power and Light Company contracts for the purchase of a 515-megawatt power reactor to be built at Oyster Creek, New Jersey. It is the first commercial power plant that is considered economical enough to be run without a federal subsidy.

1964 President Lyndon B. Johnson signs the Private Ownership of Special Nuclear Materials Act, which allows the power industry to own fuel for power reactors (rather than leasing it from the government). This allows a more normal market to develop.

1965 Expressing growing confidence in the nuclear market, the big reactor manufacturers General Electric, Babcock and Wilcox, Westinghouse, and Combustion Engineering offer their products at a fixed cost (plus inflation allowance). Between 1965 and 1967, utilities order fifty nuclear plants with a total of forty thousand megawatts of generating capacity.

1966 The Fermi fast breeder reactor outside of Detroit, Michigan, suffers a partial core meltdown that is concealed from the public. The safety of the reactor had been challenged by the AFL-CIO and other critics, but the U.S. Supreme Court had overturned a U.S. Court of Appeals decision that had halted construction of the reactor in 1956.

1969 India's first nuclear power plant, the Tarapur atomic power station, goes on-line. It was built by General Electric.

President Johnson signs the National Environmental Policy Act, which requires government agencies to submit environmental impact statements identifying the effects a proposed government policy might have on the environment. This brings environmental considerations into sharper focus in formulating government nuclear policy, but the requirement does not apply to the private sector.

The AEC establishes the Atomic Safety and Licensing Appeal Board, whose three members are responsible for technical review of the safety systems and procedures required for the granting of a construction permit for a nuclear facility.

1970 General Electric's Dresden 2 reactor near Chicago has a steam pipe break that destroys the safety system and releases radioactive iodine into the environment.

1971 President Richard M. Nixon announces funding for a project to develop a liquid-metal, fast breeder demonstration reactor in the United States.

By this year, twenty-two commercial nuclear power plants are in operation in the United States.

1973 President Nixon proposes the creation of the Energy Research and Development Administration (ERDA) to replace the Atomic Energy Commission.

The Organization of Petroleum Exporting Countries (OPEC) begins an oil boycott that leads to an "energy crisis" in the United States. Nuclear advocates quickly point to the expanded development of nuclear power as a way to lessen dependence on foreign energy supplies.

Utilities order forty-one new nuclear plants this year, and the Atomic Energy Commission (soon to be replaced by the Nuclear Regulatory Commission) predicts that one thousand nuclear plants will be in operation in the United States by the year 2000.

1974 The federal government releases a reactor safety study that concludes that a meltdown in a nuclear power reactor would be extremely unlikely.

1974 *(cont.)* The Energy Reorganization Act is signed by President Gerald Ford. The act replaces the Atomic Energy Commission with two new agencies, the Energy Research and Development Administration (ERDA) and the Nuclear Regulatory Commission (NRC).

Daniel Ford and Henry Kendall, representing the Union of Concerned Scientists and the Friends of the Earth, publish a report that concludes that emergency core cooling systems in nuclear reactors may not work.

1975 The NRC orders the shutdown of twenty-three nuclear reactors because of cracking in the coolant pipes.

The United States and Iran sign an agreement under which the United States will deliver eight nuclear reactors to Iran over the next ten years at a cost of $7 billion. The Islamic revolution in Iran in 1979 will call a halt to this program, however.

A fire started by a candle being used to find leaks damages the core at the Brown's Ferry Nuclear Plant in Decatur, Alabama.

1976 The sixty-one nuclear plants in the United States have a combined total capacity of 42,699 megawatts, amounting to 8.3 percent of the total electricity generated in the nation.

The Friends of the Earth and other organizations issue a report that views nuclear power not from the standpoint of safety but from a concern that the centralized political power of a nuclear infrastructure is antithetical to democratic values.

1977 President Jimmy Carter announces that the United States will postpone indefinitely its plans to reprocess spent nuclear fuel. He also proposes to terminate the Clinch River breeder reactor in Virginia.

President Carter signs the Energy Reorganization Act, which creates the Department of Energy and combines the Energy Research and Development Administration and the Federal Energy Administration.

The Carolina Environmental Study Group, a group of citizen activists, enlists the aid of the Public Citizen Litigation Group to challenge the Price-Anderson Act as being unconstitutional. In June, a federal district court agrees, saying that the law violates the rights of victims of nuclear accidents by denying them the just compensation required under the Fifth Amendment. (See *Carolina Environmental Study Group v. Atomic Energy Commission,* 431 F. Supp. 230, W.D.N.C. 1977.)

1978 The U.S. Supreme Court overturns the federal district court decision of the previous year, preserving the nuclear industry's limited liability.

After this year no new nuclear plants have been ordered in the United States.

1979 A stuck valve and confusion on the part of operators lead to an accident at the Three Mile Island nuclear power plant near Harrisburg, Pennsylvania. It is the most serious accident to a commercial nuclear plant in U.S. history. Although no one is directly injured by the accident, radioactive material is released into the atmosphere.

1980 President Carter presents the 1981 budget, which includes an appropriation of $1.26 billion for nonmilitary nuclear research.

France agrees to supply Iraq with weapons-grade uranium and a nuclear reactor.

The Nuclear Safety Research, Development, and Demonstration Act establishes a Department of Energy program to improve the safety of nuclear power plants through investigation and the setting of standards for the manufacture of plant components.

1981 President Ronald Reagan submits the Nuclear Power Policy Statement, which gives nuclear power a prominent role in the supplying of the nation's future energy needs.

1982 The Shippingport (Pennsylvania) nuclear power station is shut down and decommissioned after twenty-five years of operation. The decommissioning process will be completed in 1989.

1982 *(cont.)* The Nuclear Waste Policy Act (NWPA) begins a process of environmental assessment that is supposed to result in the choice of a site for a permanent, national, high-level waste storage facility. The process will encounter numerous setbacks.

1983 The Senate refuses further funding of the Clinch River breeder reactor, effectively closing down the project.

1984 The Department of Energy establishes the Civilian Radioactive Waste Management Office.

1986 The world's worst nuclear power plant accident takes place at the Soviet nuclear facility at Chernobyl in the Ukraine. An explosion destroys the plant, and thirty-five plant workers are killed. A large amount of radiation is released into the environment, about a million times what had been released in the Three Mile Island accident. There are more than two hundred cases of serious radiation sickness, and exposure is likely to cause dozens of cases of leukemia and other cancers in the future.

The Federal Emergency Management Agency (FEMA) issues a report that concludes that the plans for emergency evacuation for the area around the Pilgrim, Massachusetts, nuclear plant are inadequate.

The Nuclear Regulatory Commission reports that 430 nuclear plant shutdowns due to emergency have taken place.

The NRC cites 492 nuclear safety violations at U.S. power reactors.

At the Surry nuclear plant in Virginia, an eroded pipe carrying superheated water bursts, killing four workers. This is the only nuclear accident in the United States that has caused direct fatalities.

1987 Over one hundred nuclear plants in thirty-five states provide 12.6 percent of net U.S. energy production.

1988 The Department of Energy estimates that fifteen U.S. nuclear power plants will reach the end of their useful life span of 30–40 years by the year 2000.

The DOE awards the Bechtel Group, Inc., a large engineering and mining firm, a $1 billion contract to develop a system to transport and store radioactive nuclear wastes.

1989 Experts probing the ruined Three Mile Island reactor learn that the accident was worse than they had thought: The fuel had partially melted, but the containment building had somehow remained intact.

1990 An almost-completed nuclear power station in Midland, Michigan, is converted to burning natural gas after its owners are convinced that widespread political opposition and economic factors make nuclear operation too expensive.

1991 Japan begins construction of an advanced boiling-water reactor (ABWR). A consortium of General Electric, Hitachi, and Toshiba successfully markets the new design in Asia.

The International Commission on Radiation Protection accepts the theory that any dose of radiation, however small, is harmful.

1994 Princeton University's donut-shaped Tokamak fusion reactor produces 10.7 megawatts of power, setting a new record.

At the end of the year, 424 nuclear reactors are operating worldwide. The fastest growth is in Asia, where fourteen reactors are under construction. Japan has forty-eight nuclear plants, producing 30 percent of its electrical energy.

1995 Nineteen utilities and forty states and regulators file suit against the federal government for its likely failure to meet a 1998 congressional deadline for setting up a temporary nuclear waste storage facility.

1996 By the end of this year, six nuclear power plants close in the United States because they are no longer economical to operate. Meanwhile, Canada closes twenty-one nuclear plants for safety reasons.

1997 The Earth Summit at Kyoto, Japan, includes a proposal for nations to reduce their greenhouse gas emis-

1997 *(cont.)* sions 7 percent below 1990 levels by 2012. Nuclear power advocates suggest that expanding use of non-burning nuclear power is a key to reducing the need for "dirty" fossil fuels.

President Bill Clinton approves the establishment of the Waste Isolation Pilot Project (WIPP) in New Mexico. Its purpose is to test techniques for storing nuclear wastes so that they are sealed away from the environment.

1998 The surviving Three Mile Island nuclear plant unit 1 and the Pilgrim nuclear power station are sold to private investors. Supporters of nuclear power say this demonstrates the continuing business viability of nuclear power, whereas critics see it as an attempt to "dump" assets that utilities believe will not be cost-effective in the future.

The DOE's Energy Information Administration releases a study estimating that U.S. energy consumption will rise by 20 percent through 2020. The Clinton administration believes that some new nuclear power plants will need to be built worldwide and that the licenses for most existing plants will have to be extended.

The DOE announces the Nuclear Energy Research Initiative (NERI), a $19 million program to fund long-term studies that seek to prevent nuclear proliferation and help develop cheaper, safer reactors that generate less waste.

Duke Power and Baltimore Gas and Electric file for an extension of the operating license for their nuclear plants. The process, which requires demonstrating the safety of aging reactor vessels and internal components, is expected to cost between $15 million and $20 million.

The Nuclear Regulatory Commission certifies several advanced nuclear power plant designs, allowing builders of new plants to obtain building and operating licenses in a single step before beginning construction.

The Union of Concerned Scientists reports that their fifteen-month study of ten nuclear plants reveals careless inspections, numerous worker errors, and faulty procedures.

The U.S. Department of Energy's plan to dispose of nuclear wastes in a large underground facility at Yucca Mountain, Nevada, remains stalled, and DOE refuses to accept deliveries of spent nuclear fuel, which remains stored at nuclear plant sites.

The ruling Social Democrat–Green Party coalition in Germany announces plans to phase out all nuclear power generation in Germany.

Japan continues plans to build a commercial fast breeder nuclear reactor, and Eskom, a South African company, continues plans to build and market modular 100-kilowatt nuclear generation plants.

1999 An Associated Press poll reports that support for nuclear power has dropped to 45 percent, down 10 percent from its 1989 level.

The Department of Energy designates six nuclear power plants in the Carolinas and Virginia to generate energy from surplus plutonium from nuclear weapons stockpiles.

Workers at a Japanese uranium processing plant accidentally set off a runaway nuclear reaction. Two of the three workers will die from the effects of massive radiation exposure.

A U.S. federal appeals court reinstates two thousand lawsuits claiming health damages from the Three Mile Island accident. The judge rules that the lower court had made an error in dismissing the suits on the basis of a sample of ten cases.

The one remaining reactor at the Chernobyl site is shut down due to a small leak in a pipe. It had been restarted after months of repairs.

Despite agreeing that the nuclear industry had made deceptive ad claims about the environmental benefits of nuclear power, the Federal Trade Commission decides not to ban the ads.

2000 Steam generation tubes rupture at the Indian Point 2 Nuclear Reactor in Buchanan, New York. Although only a small amount of radioactive steam is released into the atmosphere, critics such as Ralph Nader's Public Citizen group warn that the rupture of similar tubes in aging nuclear power plants could cause disastrous core meltdowns.

A controlled burn near Los Alamos, New Mexico, erupts into a huge wildfire. Although the fire roars through portions of the Los Alamos National Laboratory facilities, officials insist that nuclear weapons materials and waste were never in danger of release.

A thirteen-year study by the University of Pittsburgh finds no increased incidence of cancer in people living near the Three Mile Island nuclear plant.

President Clinton vetoes congressional legislation that would have required the federal government to begin accepting nuclear waste at a temporary location in Nevada, pending completion of the permanent waste facility at Yucca Mountain.

German chancellor Gerhard Schroeder and the nuclear industry agree to phase out the country's nuclear power plants over a twenty-year period.

3

Biographical Sketches

Even brief biographies of all the people who played key roles in the discovery of the structure of the atom and, in particular, the process of nuclear fission would fill a large book. Because this book is focused on the application of nuclear energy to power generation, the selection has been limited to individuals who played important roles in the discovery of fission and the development of nuclear power, or who have offered criticism or opposition to its growth. Occasional references to biographies or important works written by the people discussed also are included.

Bernard Baruch (1870–1965)

Wealthy as the result of successful stock speculation, Baruch played an important role in national politics starting with his chairmanship of the War Industries Board during World War I. In 1919, he served as an economic advisor on President Woodrow Wilson's team at the Paris Peace Conference. During the 1930s he served as an adviser on economic policies to President Franklin D. Roosevelt.

When the control of nuclear energy became a key issue in the aftermath of the development of the atomic bomb during World War II, Baruch played a significant role in the attempt to create a workable international framework that would both restrain the development of nuclear weapons by other nations and promote the peaceful development of nuclear power. What became known as the Baruch Plan for International Control of Atomic Energy suggested that the United Nations impose sanctions on countries that violated agreements on the use of atomic energy, while at the same time the United States would offer help in establishing nuclear power industries in cooperating nations.

Baruch's plan met with little acceptance. Most Americans wanted the United States to jealously guard its nuclear monopoly so that the Soviets could not develop nuclear weapons to counter those of the United States. The Soviets in turn opposed the Baruch Plan, claiming that its real purpose was to perpetuate the U.S. nuclear monopoly. By 1949 the Soviets had exploded their first atomic bomb and the nuclear arms race was under way.

Baruch's autobiography, *Baruch,* was published in two volumes (Holt, 1957–1960).

Hans Albrecht Bethe (1906–)

Born in Strasbourg, France, and trained in Germany, physicist Hans Bethe went to England and then the United States following the rise of the Nazi party to power during the 1930s (Bethe's mother was Jewish). During the war he worked on the development of both radar and the atomic bomb. (When some scientists worried that an atomic explosion might set off a runaway fusion reaction in the earth's atmosphere, Bethe's calculations helped dispel their fears.) His enduring legacy, however, is in advancing the understanding of the nucleus and nuclear reactions. In particular, he was able to determine the precise nature of the nuclear fusion reaction that powers stars like our sun, in which hydrogen is the fuel but carbon serves as a catalyst. He won the Nobel Prize in physics in 1967 for this work.

In the social and political realm, Bethe has worked tirelessly to promote nuclear disarmament, a ban on nuclear weapons tests, and initiatives for the peaceful use of atomic energy. He opposed Edward Teller's plans for the development of the hydrogen bomb, hoping that both the United States and the USSR could agree to not build such superweapons. During the 1980s, when Teller promoted the Strategic Defense Initiative (also known as "Star Wars"), Bethe opposed the effort as technically impractical and politically destabilizing.

If practical fusion power is ever developed, Bethe's explanation of how the sun works will be one of the key steps that made it possible.

Niels Bohr (1885–1962)

After Bohr earned his Ph.D. in physics in 1911 from the University of Copenhagen, Denmark, he became an assistant to James J. Thompson (discoverer of the electron) at Cambridge University, England, and then worked at Manchester University under Ernest Rutherford, where he helped him make his key discoveries about the nature of the atomic nucleus and the results of bombarding it.

Bohr's key contribution, however, was in developing the comprehensive theoretical understanding of the atom that would explain Rutherford's discoveries. In 1913 Bohr used the new quantum theory to explain the orbits of electrons around the nucleus; he is, in fact, considered to be the leading figure in early quantum physics. In 1922, Bohr won the Nobel Prize in physics for his work.

During 1938 and 1939 Bohr increasingly turned his attention to the possibility of liberating vast amounts of energy through nuclear fission, which was emerging from the work of Enrico Fermi, Otto Hahn, and Lise Meitner. He came to the United States and warned scientists that Germany's Nazi party might be developing atomic weapons. In 1943 the British air force helped Bohr escape from Nazi-occupied Denmark. He joined the Manhattan Project at Los Alamos, New Mexico, where he provided theoretical help for the atom bomb effort. After the war, Bohr returned to Denmark and devoted himself to peaceful applications of atomic energy.

A good source for reading about Bohr's life and work is *Niels Bohr: His Life and Work as Seen by His Friends and Colleagues* (North-Holland, 1967).

Helen B. Caldicott (1938–)

Australian physician and antinuclear activist Helen Caldicott has brought intense energy and passion to the antinuclear movement. Her first involvement came in 1971 when she warned the public of dangerous fallout from French nuclear bomb tests in the South Pacific. The publicity helped lead to the election of an antinuclear Labor Party slate to the Australian parliament, but the movement was unsuccessful in banning exports of uranium from Australia (one of the world's major producers of the substance).

Caldicott moved to the United States in 1975 and continued her activism. Her book *Nuclear Madness: What You Can Do* (1978; revised ed., Norton, 1994) became influential in the growing public opposition to nuclear power and nuclear weapons. The same year she revived the organization Physicians for Social Responsibility and gave it a new antinuclear focus. She also founded the group Women's Action for Nuclear Disarmament in 1980. During the 1980s she played a major role in the revived opposition to nuclear weapons, as well as promoting general environmental concerns.

Her autobiography is entitled *A Desperate Passion* (Norton, 1996).

Jimmy Carter (1924–)

The thirty-ninth president of the United States, James Earl Carter was born in Plains, Georgia. Graduating from the Naval Academy at Annapolis in 1946, Carter later served aboard nuclear submarines and helped the outspoken Admiral Hyman Rickover in the navy's nuclear engineering program.

Carter's involvement with nuclear propulsion as a naval officer gave him more technical background in nuclear matters than most politicians. However, as president (1977–1981) Carter promoted an environmental agenda that minimized nuclear power, despite serious economic problems caused by the high price of oil.

In foreign affairs, he completed negotiation of the SALT II nuclear weapons limitation treaty, although its implementation was derailed following the Soviet invasion of Afghanistan. After he left office, Carter became an internationally respected and effective negotiator seeking to resolve conflicts and promote human rights.

Bernard Cohen (1924–)

Emeritus professor of physics and environmental and occupational health at the University of Pittsburgh, Bernard Leonard Cohen is a prolific writer on the health effects of radiation and has written many scientific papers on risks associated with nuclear materials. A former president of the Health Physics Society, Cohen has been an outspoken advocate of nuclear energy, asserting its fundamental safety and challenging the theory that even tiny doses of radiation necessarily create incremental cancer risks. His book *The Nuclear Energy Option: Alternatives for the 90s* (Plenum Press, 1990) is an updated version of his arguments for nuclear energy as a safe alternative to polluting fossil fuels.

Barry Commoner (1917–)

One of the most influential American environmentalists during the 1970s, Commoner, an accomplished biologist, became a leading spokesperson against nuclear power. He concluded that the combination of inherent risks in the technology and influence of the industry over those who sought to regulate it makes nuclear power unacceptable. He founded the Citizens' Party and ran for president on its ticket in 1980.

Commoner wrote two influential books in the 1970s that helped form the modern environmentalist critique of energy policy and the movement for energy alternatives: *Alternative Technologies for Power Production* (Macmillan, 1975) and *Social Costs of Power Production* (Macmillan, 1975). His book *The Poverty of Power: Energy and the Economic Crisis* (Knopf, 1976) took a broad perspective on energy economics and energy choices for the future.

Dwight D. Eisenhower (1890–1969)

Eisenhower, the chief commander of Allied forces in western Europe during World War II, emerged from the war as a popular hero and desirable political candidate. In 1952, he was elected U.S. president, serving two terms.

Eisenhower's presidency coincided with the formative years of nuclear power in the United States, and he was a forceful advocate of its development. His "Atoms for Peace" speech was the keynote for the effort to turn the atomic energy that had first shown its face as a devastating weapon of war into a bountiful and reliable source of energy for America's booming economy. Eisenhower also sought to have nations join together to share and control nuclear technology. In 1957 the International Atomic Energy Agency was created to implement these ideas, and sixty-two nations signed the agency's charter. Domestically, Eisenhower's administration was marked by the establishment of the Atomic Energy Commission as both regulator and promoter of nuclear power in the United States.

Many works have been written by or about Dwight Eisenhower. One good biography is *Eisenhower* by Stephen Ambrose, in two volumes (Simon and Schuster, 1983–1984).

Enrico Fermi (1901–1954)

Italian-born physicist Enrico Fermi, besides making important contributions to quantum physics, was a key researcher into the nature of radioactivity and atomic decay during the 1930s. He discovered element 93, neptunium, one of a series of transuranic elements that would play an important role in nuclear energy.

In 1938, Fermi went to Stockholm, Sweden, to receive the Nobel Prize for his work with radioactivity. Because of the dominance of Italy by Benito Mussolini's Fascists, however, he emigrated to the United States rather than returning home. In 1942 he led the effort to build the world's first nuclear reactor beneath the football stands at the University of Chicago. He then worked on the atomic bomb at Los Alamos, New Mexico, and on the hydrogen bomb. Fermi later played a key role serving on the General Advisory Committee of the Atomic Energy Commission, and in 1954 he received its first special prize, which was given his name. The prize is for scientific and technical achievement in the development, use, and control of atomic energy.

Emilio Segré's biography *Enrico Fermi: Physicist* (University of Chicago Press, 1970) is a good introduction to Fermi's work and its significance.

John Gofman (1918–)

John Gofman's training as both a doctor of medicine and a Ph.D. in nuclear chemistry well equipped him to play a pivotal role in establishing the modern science of health physics, which studies the effects of radiation on human health and develops ways to assess and reduce radiation exposure.

During World War II Gofman worked on the Manhattan Project and with Glenn Seaborg to develop two different methods for separating plutonium from irradiated fuel. This would lead to the development of the "plutonium factory" at the Hanford Nuclear Reservation in Washington State.

Gofman then returned to the academic world, where he had a long career as a professor of medical physics. He also served as director or administrator in the Lawrence Berkeley and Lawrence

Livermore laboratories. Meanwhile, he continued his medical research. In addition to studying radiation effects, he did important work on lipoproteins and artery and heart disease.

Gofman's work with radiation led him to be increasingly critical of what he saw as lax standards and a lack of commitment of the nuclear establishment to protecting workers and the public from the dangers of radiation. He warned that the emerging medical consensus was that "there was no safe dose" of radiation, and that it was important to avoid any unnecessary exposure. Gofman's research was attacked as "slipshod" by some critics, and his relationship with the AEC became strained. The agency cut off his research funding, but he found support for his work at Berkeley.

Gofman has testified in court on behalf of radiation victims in cases such as *Johnston v. U.S.* and *Silkwood v. Kerr-McGee.* A transcript of Gofman's oral history, "Human Radiation Studies: Remembering the Early Years," can be found on the Web at http://tis.eh.doe.gov/ohre/roadmap/histories/0457/0457toc.html#Short.

Bourke B. Hickenlooper (1896–1971)

A Republican senator from Iowa, Hickenlooper was the first chairman of the Joint Committee on Atomic Energy, which had been created by the Atomic Energy Act of 1946. This committee had oversight of the Atomic Energy Commission, the principal nuclear regulatory agency. Hickenlooper was particularly interested in promoting the development of nuclear power in countries friendly to the United States. To promote this effort, he cosponsored a bill in 1954 that eased restrictions on the sharing of "special nuclear materials" with friendly nations.

Henry M. Jackson (1912–1983)

Jackson, a Democratic U.S. senator from the state of Washington, had a long and distinguished career that spanned issues from national defense to labor and civil rights legislation. He served in the House of Representatives starting in 1940 and was elected to the Senate in 1953.

In 1955, Jackson proposed that President Eisenhower create a "bank" of nuclear materials not needed for weapons production that could be used for both military and civilian purposes. Eisenhower turned down the idea, however.

Jackson naturally took a special interest in the huge Hanford, Washington, nuclear production facility. He gave his backing to the development of a dual-purpose nuclear reactor design that could produce plutonium for both weapons and electric power. Throughout his career Jackson remained a strong proponent of expansion of nuclear power, including its use for generating electricity in remote areas such as Antarctica.

David E. Lilienthal (1899–1981)

An Illinois lawyer, Lilienthal became involved in the power industry when Governor Philip La Follette appointed him to the Wisconsin Public Service Commission. Later, President Roosevelt appointed him as one of three directors of the huge Tennessee Valley Authority (TVA). Lilienthal's insistence on separating the agency from politics earned him the enmity of some industrial interests.

In 1946 President Truman appointed Lilienthal as the first chairman of the Atomic Energy Commission, where he was largely responsible for defining the role and implementing the operations of the new agency within the guidelines set by the Atomic Energy Act. As he set to work to organize the agency and its key program for the development of nuclear power reactors, he tried to listen to public concerns and make the agency accountable. However, the growing Cold War fear of Soviet espionage and nuclear weapons led to an investigation of the AEC's security by Senator Bourke B. Hickenlooper and charges of mismanagement, which led to Lilienthal's resignation in 1949.

Lilienthal remained a thoughtful advocate of nuclear energy. His views can be found in his book *Atomic Energy: A New Start* (Harper and Row, 1980).

Lise Meitner (1878–1968)

Born in Vienna, Austria, Meitner obtained her doctorate in physics from Vienna University in 1906, then went to Berlin, Germany, and began a long-term collaboration with chemist Otto Hahn. In 1918 they discovered the element protactinium, and Meitner also did important research on radioactive decay. Despite the pervasive prejudice that erected formidable obstacles to women pursuing a career in science, Meitner persevered, becoming professor of physics at the University of Berlin in 1926. Adolf Hitler's rise to power, however, put increasing pressure on Meitner and other Jewish scientists in Germany, who were gradually stripped of their professional status.

Meitner's most important work came in the late 1930s, when she closely followed Hahn's work on the bombardment of heavy elements such as uranium. Studying these results, she came to the conclusion that uranium could fission, or split apart into two approximately equal pieces, and wrote the first paper describing the phenomenon.

Although she played no part in the development of the atomic bomb and little part in the subsequent development of nuclear power, Meitner's importance has been recognized in recent years. She did not receive a Nobel Prize (possibly because she was a woman and had worked in the background) but did receive many other honors. In 1997 chemical element number 109 was named meitnerium in her honor.

A good biography is *Lise Meitner: A Life in Physics* by Ruth L. Sime (University of California Press, 1996).

G. C. Minor (1937–1999)

Gregory Charles Minor was one of three managing engineers who resigned from the General Electric Reactor Division in 1976 to protest what they saw as disregard for public safety in the operation of nuclear power plants. He and his two colleagues, Richard B. Hubbard and Dale G. Bridenbaugh, testified before Congress, arguing that there was insufficient oversight and coordination in managing nuclear plants, so that serious risks were not being detected or addressed. They also founded a consulting firm, MHB Technical Associates, which conducted evaluations and studies of nuclear safety, and served as technical advisers to the controversial movie *The China Syndrome,* which depicted a disastrous nuclear meltdown.

Mayumi Oda (1941–)

Japanese American silkscreen artist Mayumi Oda has brought together art, Zen Buddhism, peace activism, and strong feminine archetypes in her much-acclaimed work. During the 1960s, married to an American scientist and living in Cambridge, Massachusetts, Oda became involved with the growing women's movement. She then learned that Japan was beginning an ambitious program of building breeder reactors to create fuel for a burgeoning nuclear power industry. For a time, she set her art aside in favor of antinuclear activism, helping to found the group Plutonium-Free Future. She feels that the spread of nuclear power in Japan and throughout Asia might well fuel a nuclear arms race, and she argues that declining prices for uranium now make it economically unnecessary to produce plutonium fuel in breeder reactors. She also suggests that it is time for Asia and the world to turn to alternative forms of energy. Today she continues her efforts to educate people about nuclear issues and has also resumed her artistic career.

J. Robert Oppenheimer (1904–1967)

Oppenheimer was born in New York City but received his Ph.D. in physics from the University of Göttingen, Germany, in 1927. He performed important early research in quantum theory and the theory of antiparticles (particles, such as the positron, that are oppositely charged or oppositely spun from their normal counterparts).

From 1942 to 1945 Oppenheimer was director of the atomic bomb project at Los Alamos, New Mexico. Although he made key contributions to developing nuclear weapons, his experience of their devastating effects spurred him to oppose development of the more powerful hydrogen bomb and to advocate putting nuclear weapons under international control. From 1945 to 1952 Oppenheimer was chairman of the General Advisory Committee of the Atomic Energy Agency as well as a leading consultant to the U.S. delegation to the United Nations Atomic Energy Committee.

Oppenheimer's liberal views and suspected left-wing connections made him a target of the anticommunist forces of the McCarthy era. In 1953, he was suspended by the AEC as a security risk. He received much support from his academic colleagues, however, and in 1954 he was unanimously reelected as the director of the Institute for Advanced Study at Princeton.

Readers interested in Oppenheimer's life and work have a variety of choices. One of the best is *J. Robert Oppenheimer: Shatterer of Worlds* by Peter Goodchild (Houghton Mifflin, 1981). This book was produced in conjunction with the BBC/WBGH television series *Oppenheimer.*

Dixy Lee Ray (1914–1994)

After a long career as a marine biologist, Ray became the first female chair of the Atomic Energy Commission, serving from 1973 to 1975. She was a staunch advocate of nuclear power and had little patience for its critics. (She once pointed out that three hundred people a year die just from choking on food, far more than have died in decades of nuclear power in the United States.) However, Ray was well aware of safety concerns, and she campaigned to eliminate defects in the construction and operation of nuclear power plants.

When the AEC was abolished in 1974 (to be replaced by the Nuclear Regulatory Commission), the nation was in the midst of the "energy crisis" that led many Americans to think seriously for the first time about energy alternatives and choices. Nuclear power, too, was about to suffer the shock of Three Mile Island and come under a firestorm of criticism. Leaving the federal government, Ray served as governor of Washington State from 1977 to 1981, having run as an independent. The idiosyncratic Ray bucked the tide of popular opinion in refusing to work to close the large nuclear production and waste facility at Hanford, Washington. She later became an independent consultant and wrote two books that clash head-on with environmentalists: *Environmental Overkill* (HarperPerennial, 1994) and *Trashing the Planet* (Regnery Gateway, 1990).

Hyman G. Rickover (1900–1987)

Admiral Rickover played a key role in developing nuclear power as a source of propulsion for the U.S. Navy's submarines. He argued that with their ability to travel submerged indefinitely, such ships would be an essential tool for controlling the seas in future wars. His efforts culminated with the launching of the first nuclear submarine, *Nautilus,* in 1954. Rickover later became chief of the Naval Reactors Branch of the Atomic Energy Commission.

Rickover's outspoken, even badgering style often involved him in controversy and acrimony, but his determination led to the triumph of the "nuclear navy" against entrenched, old-guard opposition. Aircraft carriers and cruisers were also built with nuclear reactors. The success of the naval reactors proved to be an important spur to the development of civilian nuclear power, serving as proof of the concept that reliable reactors could be built and run for long periods without mishap.

In his later years Rickover became interested in educational reform and wrote several books on the subject. He received the Enrico Fermi Award in 1965 and the Medal of Freedom in 1980.

Glenn T. Seaborg (1912–1999)

Chemist Glenn T. Seaborg developed much of the understanding of the properties of transuranic elements (such as plutonium) that was necessary for the development of nuclear power. In 1951 he shared a Nobel Prize in chemistry with Edwin McMillan for this work.

In 1961, President Kennedy appointed Seaborg as chairman of the Atomic Energy Commission. He served ten years, the longest term for any AEC chairman. Seaborg believed that nuclear power was both technically and economically viable. He was an effective leader, combining the qualities of a scientist and a good administrator. He received the Enrico Fermi Award in 1959. In 1997 chemical element number 106 was designated seaborgium in his honor.

Seaborg's book *The Nuclear Milestones* (W. H. Freeman, 1972) offers a good overview of atomic discoveries and the development of nuclear power.

Karen Silkwood (1946–1974)

Karen Silkwood became a cause célèbre of the antinuclear movement when she died in a car crash on November 13, 1974. She had been working as a chemical technician in the Kerr-McGee plutonium production plant at Crescent, Oklahoma. Prior to her death, she had been helping the Oil, Chemical and Atomic Workers' Union gather evidence about alleged unsafe conditions at the plant. She was exposed to plutonium under circumstances that remain mysterious. After her death, her body was examined and plutonium was found in her tissues. This suggested to nuclear opponents that she had been deliberately poisoned with the radioactive substance.

After Silkwood's death, her estate filed a lawsuit against Kerr-McGee, charging the company with maintaining unsafe working conditions. A jury awarded the estate $10.5 million. However, a federal appeals court reversed all but $5,000 of the award. As the estate prepared further appeals, the company settled out of court for $1.3 million. The Kerr-McGee fuel plant closed in 1975.

For more about the Silkwood mystery and case, see *The Killing of Karen Silkwood: The Story Behind the Plutonium Case* by Richard L. Rashke (Houghton Mifflin, 1981).

Lewis L. Strauss (1896–1974)

Lewis L. Strauss served in a variety of administrative capacities. During World War I he served in the Belgian Relief Commission under Herbert Hoover. During World War II he served as special assistant to Secretary of the Navy James Forrestal while attaining the rank of rear admiral.

In 1953 President Eisenhower appointed Strauss as chairman of the Atomic Energy Commission. He served simultaneously as Eisenhower's special assistant for atomic energy matters. His term on the AEC (1953–1958) was controversial. He engaged in a power struggle with the congressional Joint Committee on Atomic Energy over the role of the AEC and its relationship to the committee's oversight role. He also disagreed with J. Robert Oppenheimer, who did not share his strong support for developing the hydrogen bomb.

Strauss saw the role of the AEC as being more independent of Congress but also more limited in its scope. He believed that the federal government should play only a limited role in the Power Reactor Demonstration Program, which was designed to build demonstration reactors to show the commercial feasibility of nuclear power, and to help private industry start nuclear power programs. He believed that the government should not go beyond demonstration projects to the active subsidization of the nuclear industry.

Strauss's book *Men and Decisions* (Doubleday, 1962) recounts many of the issues he dealt with during his term on the AEC.

Grace Thorpe (1924–)

The daughter of famed Native American Olympic star Jim Thorpe, Grace Thorpe became active in nuclear issues in 1992 when her tribe, the Sac and Fox Nation, accepted a $100,000 grant from the federal government to study the establishment of nuclear waste dumps on tribal land. As the tribe's health commissioner, Thorpe organized a campaign to warn her people of the dangers of nuclear waste and reminded them of the many agreements with Native Americans that the government had broken in the past. The tribe then voted to reject the government grant.

Thorpe went on in 1993 to found the National Environmental Coalition of Native Americans (NECONA) and now travels around the country, helping other tribes organize against nuclear waste. Twenty "Nuclear Free Zones" have been established on reservations, and fourteen of the seventeen tribes that had originally sought to participate in the federal nuclear waste program have withdrawn their applications.

Stewart Udall (1920–)

Stewart Lee Udall is part of a famous Arizona political family—his father, Levi S. Udall, was an Arizona Supreme Court justice, and his brother is former congressperson Morris Udall. In his youth he served as a Mormon missionary for two years and served in the U.S. Air Force as a gunner during World War II. He began his long congressional career in 1954. He then served as Secretary of the Interior under Presidents Kennedy and Johnson.

Udall's legislative interests show a major focus on protecting the environment, including preservation and expansion of parks and wilderness areas in his home state. During the 1970s, he became a strong advocate of solar energy as a solution to the energy crisis. He also took up the issue of compensation for health effects suffered by uranium miners and their families, and in 1990 helped author the Radiation Exposure Compensation Act. He continues to address this and other aspects of the Cold War's nuclear legacy, as recounted in his book *The Myths of August: A Personal Exploration of Our Tragic Cold War Affair with the Atom* (Rutgers University Press, 1998).

4

Facts, Illustrations, and Documents

This chapter presents facts, statistics, illustrations, and statements that provide background information about nuclear power and the issues surrounding nuclear plant operation and the management of nuclear waste. Readers interested in further information on any of these topics are encouraged to consult the list of websites in Chapter 7, many of which offer a veritable treasure trove of information.

Design and Operation of Nuclear Power Plants

The following tables and illustrations describe the basic designs of nuclear power plants, statistics about the role played by nuclear power in the United States and around the world, and basic information about every commercial nuclear power plant in the United States.

Basic Nuclear Plant Designs

Commercial nuclear power plants in the United States are of two types: pressurized water reactors (PWRs) and boiling water reactors (BWRs). The following text and illustrations describe the basic operation of these two types of plants.

Pressurized Water Reactor

The PWR (Figure 4.1) is a two-stage system that keeps the water in the reactor under high pressure so that it does not boil. Piping

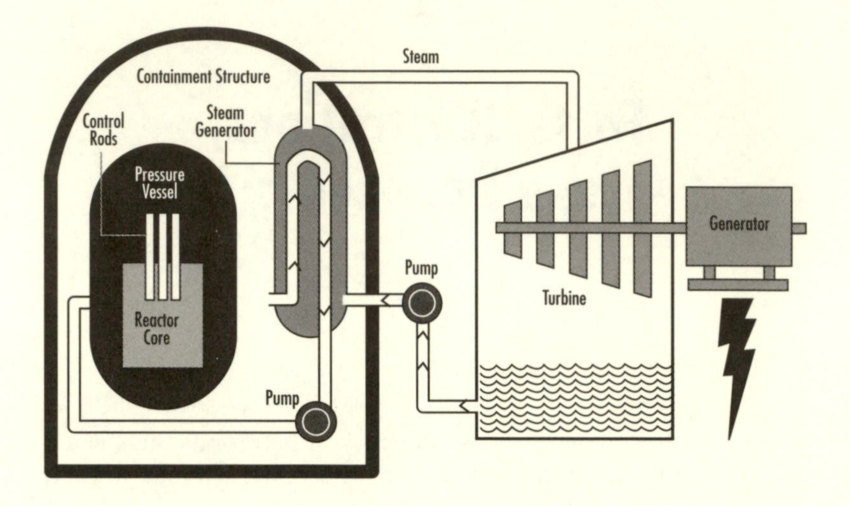

Figure 4.1 Pressurized water reactor (U.S. Nuclear Regulatory Commission, NRC: Regulator of Nuclear Power, *p. 10)*

carries this heated water to large cylinders called steam generators. The heated reactor water flows through thousands of tubes in the steam generator. The tubes are surrounded by a secondary water supply that boils and produces steam, which is carried away by pipes to spin the turbine generator.

The reactor cooling water then returns to the reactor to be reheated and circulated back to the steam generator, again in a continuous loop.

Source: U.S. Nuclear Regulatory Commission, *NRC: Regulator of Nuclear Power,* pp. 9–10.

Boiling Water Reactor

The BWR (Figure 4.2) is a single-stage system that allows the water in the reactor to boil and produce steam, which is then piped directly to the turbine generator.

In both types of reactors, once the steam loses its energy in spinning the turbine, it flows into a condenser. Because not all the heat energy in the steam can be converted into electricity, the leftover heat must be carried away by cooling water pumped through the condenser. The condenser contains thousands of tubes carrying cool water, which causes the steam to condense back to water. The water is collected in the condenser and

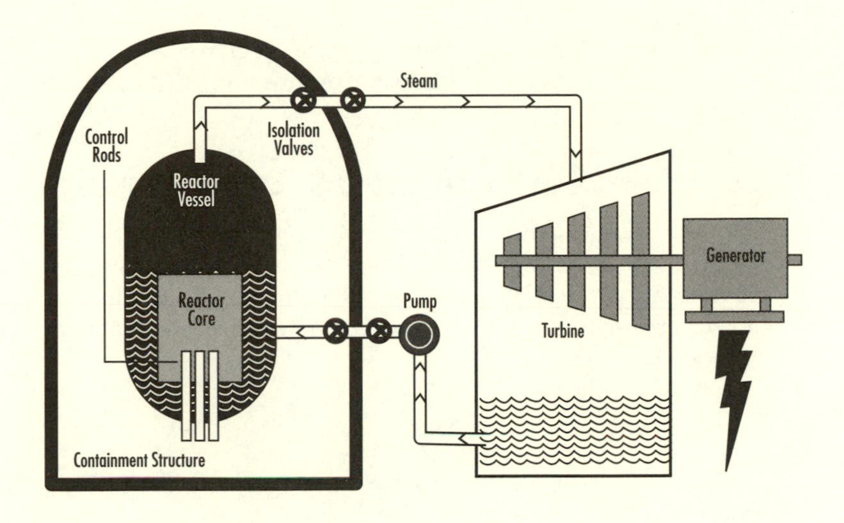

Figure 4.2 Boiling water reactor (U.S. Nuclear Regulatory Commission, NRC: Regulator of Nuclear Power, *p. 11)*

pumped back to be reheated, either in the reactor itself for a BWR or in the PWR steam generator.

The condenser cooling water is drawn from a lake, river, or ocean. This water does not actually pass through the reactor, but only through the condenser tubes to cool the steam after it goes through the turbine. There is no contact between the condenser cooling water and the reactor cooling water or the reactor components themselves.

This outside cooling water, heated as it passes through the condenser, is returned to the source lake, river, or ocean. Many nuclear plants as well as other types of power plants or industrial facilities use cooling towers, cooling lakes or ponds, or other techniques to reduce the effects of heated water discharged directly back into the river, lake, or ocean. With cooling towers the water, once cooled, is usually pumped back into the condenser to be used again to carry off heat. Only a small portion of the water is returned to the natural water source.

Once the steam is produced, the process and equipment used to generate electricity are similar in nuclear power plants and fossil-fueled power plants. Fossil-fueled plants also discharge heated water to the environment or use cooling towers or other means to cool the water from the condenser.

Source: U.S. Nuclear Regulatory Commission, *NRC: Regulator of Nuclear Power,* pp. 9–10.

Nuclear Contribution to the U.S. Energy Supply

The percentage of U.S. electric power generated by nuclear power plants has been slowly declining but is still significant, especially because nuclear plants could increase production by about 50 percent if a crisis were to cut off supplies of conventional fuels (Figure 4.3).

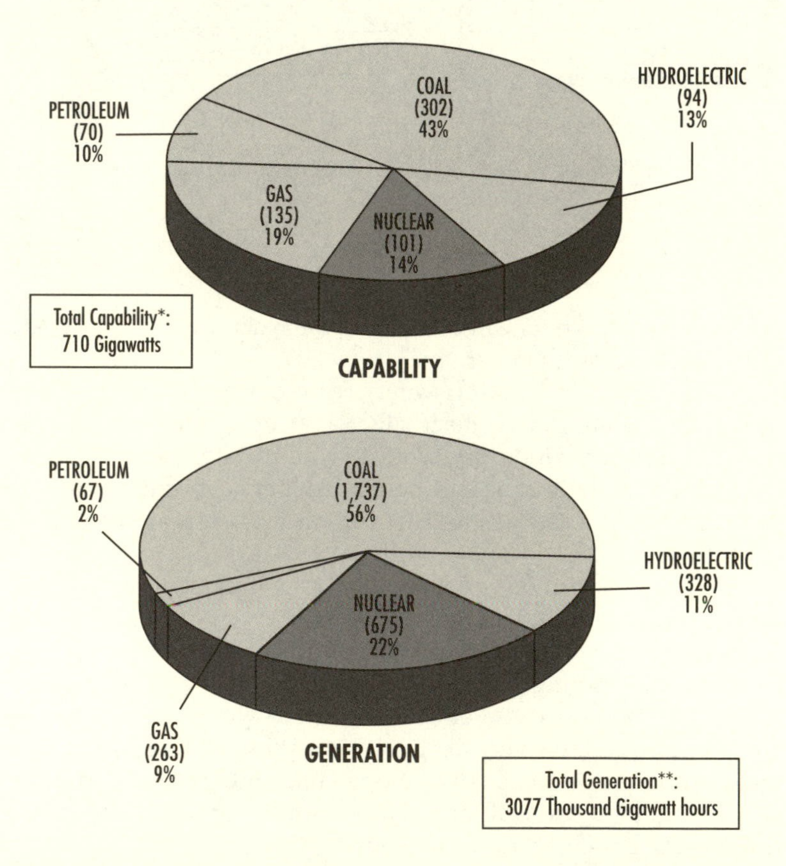

Figure 4.3 U.S. electric capability and net generation by energy source, 1996 (DOE/EIA Inventory of Power Plants in the United States as of January 1, 1997, Table 1, p. 19, and DOE/EIA Monthly Energy Review, Table 7.1, p. 95)

Reactors in the United States

Figure 4.4 shows the locations of power reactors currently in use in North America.

Table 4.1 provides basic statistics on each operating (or recently shut down) power reactor in the United States. Information includes reactor manufacturer and type, generating capacity, and brief information about the site.

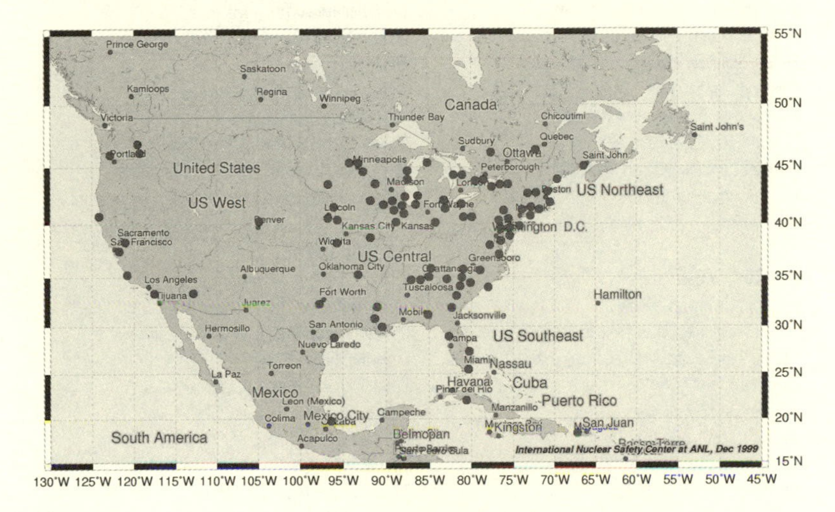

Figure 4.4 North American reactors (Argonne National Laboratory)

TABLE 4.1
Nuclear Reactors Operating in the United States, 1998

ARKANSAS NUCLEAR 1
Location: Russellville, Arkansas
Utility: Arkansas Power and Light Company
Reactor Supplier: Babcock and Wilcox
Capacity: 836 net MWe
Reactor Type: Pressurized water reactor
Date of Operation: May 1974
License Expiration Date: 05/20/2014
Electricity Produced in 1998: 6.2 billion kWh
1998 Average Capacity Factor: 84.9%

ARKANSAS NUCLEAR 2
Location: Russellville, Arkansas
Utility: Arkansas Power and Light Company
Reactor Supplier: Combustion Engineering, Inc.
Capacity: 858 net MWe
Reactor Type: Pressurized water reactor
Date of Operation: September 1978
License Expiration Date: 07/17/2018
Electricity Produced in 1998: 6.9 billion kWh
1998 Average Capacity Factor: 91.6%

BEAVER VALLEY 1
Location: Shippingport, Pennsylvania
Utility: Duquesne Light Company
Reactor Supplier: Westinghouse Corporation
Capacity: 810 net MWe
Reactor Type: Pressurized water reactor

(continues)

TABLE 4.1
(continued)

Date of Operation: July 1976
License Expiration Date: 01/29/2016
Electricity Produced in 1998: 2.8 billion kWh
1998 Average Capacity Factor: 39.2%

BEAVER VALLEY 2
Location: Shippingport, Pennsylvania
Utility: Duquesne Light Company
Reactor Supplier: Westinghouse Corporation
Capacity: 820 net MWe
Reactor Type: Pressurized water reactor
Date of Operation: August 1987
License Expiration Date: 05/27/2027
Electricity Produced in 1998: 1.7 billion kWh
1998 Average Capacity Factor: 23.7%

BIG ROCK POINT (retired)
Location: Charlevoix, Michigan
Utility: Consumers Power Company
Reactor Supplier: Westinghouse Corporation
Capacity: 67 net MWe
Reactor Type: Boiling water reactor
Date of Operation: August 1962
License Expiration Date: 05/31/2000
Electricity Produced in 1998: 0.0 billion kWh
1998 Average Capacity Factor: 0.0%

BRAIDWOOD 1
Location: Braidwood, Illinois
Utility: Commonwealth Edison Company
Reactor Supplier: Westinghouse Corporation
Capacity: 1090 net MWe
Reactor Type: Pressurized water reactor
Date of Operation: July 1987
License Expiration Date: 10/17/2026
Electricity Produced in 1998: 7.6 billion kWh
1998 Average Capacity Factor: 79.3%

BRAIDWOOD 2
Location: Braidwood, Illinois
Utility: Commonwealth Edison Company
Reactor Supplier: Westinghouse Corporation
Capacity: 1090 net MWe
Reactor Type: Pressurized water reactor
Date of Operation: May 1988
License Expiration Date: 12/18/2027
Electricity Produced in 1998: 9.7 billion kWh
1998 Average Capacity Factor: 101.5%[a]

BROWN'S FERRY 1
Location: Decatur, Alabama
Utility:Tennessee Valley Authority
Reactor Supplier: General Electric Company
Capacity:1065 net MWe
Reactor Type: Boiling water reactor
Date of Operation: December 1973
License Expiration Date: 12/20/2013
Electricity Produced in 1998: 0.0 billion kWh
1998 Average Capacity Factor: 0.0%

BROWN'S FERRY 2
Location: Decatur, Alabama
Utility:Tennessee Valley Authority
Reactor Supplier: General Electric Company
Capacity:1065 net MWe
Reactor Type: Boiling water reactor
Date of Operation: August 1974
License Expiration Date: 06/28/2014
Electricity Produced in 1998: 9.3 billion kWh
1998 Average Capacity Factor: 99.7%

BROWN'S FERRY 3
Location: Decatur, Alabama
Utility: Tennessee Valley Authority
Reactor Supplier: General Electric Company
Capacity: 1065 net MWe
Reactor Type: Boiling water reactor
Date of Operation: August 1976
License Expiration Date: 07/02/2016
Electricity Produced in 1998: 7.9 billion kWh
1998 Average Capacity Factor: 84.5%

BRUNSWICK 1
Location: Southport, North Carolina
Utility: Carolina Power and Light Company
Reactor Supplier: General Electric Company
Capacity: 820 net MWe
Reactor Type: Boiling water reactor
Date of Operation: November 1976
License Expiration Date: 09/08/2016

(continues)

TABLE 4.1
(continued)

Electricity Produced in 1998: 6.4 billion kWh
1998 Average Capacity Factor: 88.6%

BRUNSWICK 2
Location: Southport, North Carolina
Utility: Carolina Power and Light Company
Reactor Supplier: General Electric Company
Capacity: 811 net MWe
Reactor Type: Boiling water reactor
Date of Operation: December 1974
License Expiration Date: 12/27/2014
Electricity Produced in 1998: 7.0 billion kWh
1998 Average Capacity Factor: 97.9%

BYRON 1
Location: Byron, Illinois
Utility: Commonwealth Edison Company
Reactor Supplier: Westinghouse Corporation
Capacity: 1120 net MWe
Reactor Type: Pressurized water reactor
Date of Operation: February 1985
License Expiration Date: 10/31/2024
Electricity Produced in 1998: 7.8 billion kWh
1998 Average Capacity Factor: 79.9%

BYRON 2
Location: Byron, Illinois
Utility: Commonwealth Edison Company
Reactor Supplier: Westinghouse Corporation
Capacity: 1120 net MWe
Reactor Type: Pressurized water reactor
Date of Operation: January 1987
License Expiration Date: 11/06/2026
Electricity Produced in 1998: 8.6 billion kWh
1998 Average Capacity Factor: 87.6%

CALLAWAY 1
Location: Fulton, Missouri
Utility: Union Electric Company
Reactor Supplier: Westinghouse Corporation
Capacity: 1143 net MWe
Reactor Type: Pressurized water reactor
Date of Operation: October 1984
License Expiration Date: 10/18/2024
Electricity Produced in 1998: 8.5 billion kWh
1998 Average Capacity Factor: 85.1%

CALVERT CLIFFS 1
Location: Lusby, Maryland
Utility: Baltimore Gas and Electric Company
Reactor Supplier: Combustion Engineering, Inc.
Capacity: 835 net MWe
Reactor Type: Pressurized water reactor
Date of Operation: July 1974
License Expiration Date: 07/31/2014
Electricity Produced in 1998: 6.1 billion kWh
1998 Average Capacity Factor: 83.5%

CALVERT CLIFFS 2
Location: Lusby, Maryland
Utility: Baltimore Gas and Electric Company
Reactor Supplier: Combustion Engineering, Inc.
Capacity: 840 net MWe
Reactor Type: Pressurized water reactor
Date of Operation: November 1976
License Expiration Date: 08/31/2016
Electricity Produced in 1998: 7.2 billion kWh
1998 Average Capacity Factor: 98.1%

CATAWBA 1
Location: Clover, South Carolina
Utility: Duke Power Company
Reactor Supplier: Westinghouse Corporation
Capacity: 1129 net MWe
Reactor Type: Pressurized water reactor
Date of Operation: January 1985
License Expiration Date: 12/06/2024
Electricity Produced in 1998: 8.9 billion kWh
1998 Average Capacity Factor: 90.0%

CATAWBA 2
Location: Clover, South Carolina
Utility: Duke Power Company
Reactor Supplier: Westinghouse Corporation
Capacity: 1129 net MWe
Reactor Type: Pressurized water reactor
Date of Operation: May 1986
License Expiration Date: 02/24/2026
Electricity Produced in 1998: 8.7 billion kWh
1998 Average Capacity Factor: 87.7%

CLINTON 1
Location: Clinton, Illinois

(continues)

TABLE 4.1
(continued)

Utility: Illinois Power Company
Reactor Supplier: General Electric Company
Capacity: 930 net MWe
Reactor Type: Boiling water reactor
Date of Operation: April 1987
License Expiration Date: 09/29/2026
Electricity Produced in 1998: –0.1 billion kWh[b]
1998 Average Capacity factor: 0.0%

COMANCHE PEAK 1
Location: Glen Rose, Texas
Utility: Texas Utilities Electric Company
Reactor Supplier: Westinghouse Corporation
Capacity: 1150 net MWe
Reactor Type: Pressurized water reactor
Date of Operation: April 1990
License Expiration Date: 02/08/2030
Electricity Produced in 1998: 8.5 billion kWh
1998 Average Capacity Factor: 84.4%

COMANCHE PEAK 2
Location: Glen Rose, Texas
Utility: Texas Utilities Electric Company
Reactor Supplier: Westinghouse Corporation
Capacity: 1150 net MWe
Reactor Type: Pressurized water reactor
Date of Operation: April 1993
License Expiration Date: 02/02/2033
Electricity Produced in 1998: 9.3 billion kWh
1998 Average Capacity Factor: 92.8%

COOPER 1
Location: Brownville, Nebraska
Utility: Nebraska Public Power District
Reactor Supplier: General Electric Company
Capacity: 774 net MWe
Reactor Type: Boiling water reactor
Date of Operation: January 1974
License Expiration Date: 01/18/2014
Electricity Produced in 1998: 4.9 billion kWh
1998 Average Capacity Factor: 71.8%

CRYSTAL RIVER 3
Location: Red Level, Florida
Utility: Florida Power Corporation
Reactor Supplier: Babcock & Wilcox
Capacity: 812 net MWe
Reactor Type: Pressurized water reactor
Date of Operation: January 1977
License Expiration Date: 12/03/2016
Electricity Produced in 1998: 5.9 billion kWh
1998 Average Capacity Factor: 82.6%

DAVIS BESSE 1
Location: Oak Harbor, Ohio
Utility: Toledo Edison Company
Reactor Supplier: Babcock & Wilcox
Capacity: 873 net MWe
Reactor Type: Pressurized water reactor
Date of Operation: April 1977
License Expiration Date: 04/22/2017
Electricity Produced in 1998: 6.1 billion kWh
1998 Average Capacity Factor: 80.3%

DIABLO CANYON 1
Location: Avila Beach, California
Utility: Pacific Gas and Electric Company
Reactor Supplier: Westinghouse Corporation
Capacity: 1073 net MWe
Reactor Type: Pressurized water reactor
Date of Operation: November 1984
License Expiration Date: 09/22/2021
Electricity Produced in 1998: 9.0 billion kWh
1998 Average Capacity Factor: 95.4%

DIABLO CANYON 2
Location: Avila Beach, California
Utility: Pacific Gas and Electric Company
Reactor Supplier: Westinghouse Corporation
Capacity: 1087 net MWe
Reactor Type: Pressurized water reactor
Date of Operation: August 1985
License Expiration Date: 04/26/2025
Electricity Produced in 1998: 8.1 billion kWh
1998 Average Capacity Factor: 85.5%

DONALD C. COOK 1
Location: Bridgman, Michigan
Utility: Indiana/Michigan Power Company
Reactor Supplier: Westinghouse Corporation

(continues)

TABLE 4.1
(continued)

Capacity: 1000 net MWe
Reactor Type: Pressurized water reactor
Date of Operation: October 1974
License Expiration Date: 10/25/2014
Electricity Produced in 1998: 0.0 billion kWh
1998 Average Capacity Factor: 0.0%

DONALD C. COOK 2
Location: Bridgman, Michigan
Utility: Indiana/Michigan Power Company
Reactor Supplier: Westinghouse Corporation
Capacity: 1060 net MWe
Reactor Type: Pressurized water reactor
Date of Operation: December 1977
License Expiration Date: 12/23/2017
Electricity Produced in 1998: 0.0 billion kWh
1998 Average Capacity Factor: 0.0%

DRESDEN 2
Location: Morris, Illinois
Utility: Commonwealth Edison Company
Reactor Supplier: General Electric Company
Capacity: 772 net MWe
Reactor Type: Boiling water reactor
Date of Operation: December 1969
License Expiration Date: 01/10/2006
Electricity Produced in 1998: 5.6 billion kWh
1998 Average Capacity Factor: 83.3%

DRESDEN 3
Location: Morris, Illinois
Utility: Commonwealth Edison Company
Reactor Supplier: General Electric Company
Capacity: 773 net MWe
Reactor Type: Boiling water reactor
Date of Operation: March 1971
License Expiration Date: 01/12/2011
Electricity Produced in 1998: 6.2 billion kWh
1998 Average Capacity Factor: 92.1%

DUANE ARNOLD
Location: Palo, Iowa
Utility: Iowa Electric Light and Power Company
Reactor Supplier: General Electric Company
Capacity: 535 net MWe
Reactor Type: Boiling water reactor
Date of Operation: February 1974
License Expiration Date: 02/21/2014
Electricity Produced in 1998: 3.8 billion kWh
1998 Average Capacity Factor: 80.4%

FERMI 2
Location: Newport, Michigan
Utility: Detroit Edison Company
Reactor Supplier: General Electric Company
Capacity: 1100 net MWe
Reactor Type: Boiling water reactor
Date of Operation: July 1985
License Expiration Date: 03/20/2025
Electricity Produced in 1998: 7.1 billion kWh
1998 Average Capacity Factor: 74.0%

FORT CALHOUN 1
Location: Fort Calhoun, Nebraska
Utility: Omaha Public Power District
Reactor Supplier: Combustion Engineering, Inc.
Capacity: 476 net MWe
Reactor Type: Pressurized water reactor
Date of Operation: August 1973
License Expiration Date: 08/09/2013
Electricity Produced in 1998: 3.4 billion kWh
1998 Average Capacity Factor: 81.3%

GRAND GULF 1
Location: Port Gibson, Mississippi
Utility: System Energy Resources Inc.
Reactor Supplier: General Electric Company
Capacity: 1200 net MWe
Reactor Type: Boiling water reactor
Date of Operation: November 1984
License Expiration Date: 06/16/2022
Electricity Produced in 1998: 9.2 billion kWh
1998 Average Capacity Factor: 87.4%

H. B. ROBINSON 2
Location: Hartsville, South Carolina
Utility: Carolina Power and Light Company
Reactor Supplier: Westinghouse Corporation
Capacity: 683 net MWe
Reactor Type: Pressurized water reactor

(continues)

TABLE 4.1
(continued)

Date of Operation: September 1970
License Expiration Date: 07/31/2010
Electricity Produced in 1998: 5.6 billion kWh
1998 Average Capacity Factor: 93.4%

HADDAM NECK (retired)
Location: Haddam Neck, Connecticut
Utility: Connecticut Yankee Atomic Power Company
Reactor Supplier: Westinghouse Corporation
Capacity: 560 net MWe
Reactor Type: Pressurized water reactor
Date of Operation: June 1967
License Expiration Date: 06/29/2007
Electricity Produced in 1998: 0.0 billion kWh
1998 Average Capacity Factor: 0.0%

HATCH 1
Location: Baxley, Georgia
Utility: Georgia Power Company
Reactor Supplier: General Electric Company
Capacity: 802 net MWe
Reactor Type: Boiling water reactor
Date of Operation: October 1974
License Expiration Date: 08/06/2014
Electricity Produced in 1998: 7.0 billion kWh
1998 Average Capacity Factor: 98.9%

HATCH 2
Location: Baxley, Georgia
Utility: Georgia Power Company
Reactor Supplier: General Electric Company
Capacity: 820 net MWe
Reactor Type: Boiling water reactor
Date of Operation: June 1978
License Expiration Date: 06/13/2018
Electricity Produced in 1998: 5.8 billion kWh
1998 Average Capacity Factor: 81.1%

HOPE CREEK 1
Location: Salem, New Jersey
Utility: Public Service Electric and Gas Company
Reactor Supplier: General Electric Company
Capacity: 1031 net MWe
Reactor Type: Boiling water reactor
Date of Operation: July 1986
License Expiration Date: 04/11/2026
Electricity Produced in 1998: 8.7 billion kWh
1998 Average Capacity Factor: 96.5%

INDIAN POINT 2
Location: Buchanan, New York
Utility: Consolidated Edison Company
Reactor Supplier: Westinghouse Corporation
Capacity: 931 net MWe
Reactor Type: Pressurized water reactor
Date of Operation: September 1973
License Expiration Date: 09/28/2013
Electricity Produced in 1998: 2.5 billion kWh
1998 Average Capacity Factor: 30.2%

INDIAN POINT 3
Location: Buchanan, New York
Utility: Power Authority of the State of New York
Reactor Supplier: Westinghouse Corporation
Capacity: 970 net MWe
Reactor Type: Pressurized water reactor
Date of Operation: April 1976
License Expiration Date: 12/15/2015
Electricity Produced in 1998: 7.7 billion kWh
1998 Average Capacity Factor: 90.1%

JAMES FITZPATRICK 1
Location: Scriba, New York
Utility: Power Authority of the State of New York
Reactor Supplier: General Electric Company
Capacity: 820 net MWe
Reactor Type: Boiling water reactor
Date of Operation: October 1974
License Expiration Date: 10/17/2014
Electricity Produced in 1998: 4.9 billion kWh
1998 Average Capacity Factor: 68.6%

JOSEPH M. FARLEY 1
Location: Dothan, Alabama
Utility: Alabama Power Company
Reactor Supplier: Westinghouse Corporation
Capacity: 823 net MWe
Reactor Type: Pressurized water reactor
Date of Operation: June 1977
License Expiration Date: 06/25/2017

(continues)

TABLE 4.1
(continued)

Electricity Produced in 1998: 5.2 billion kWh
1998 Average Capacity Factor: 72.5%

JOSEPH M. FARLEY 2
Location: Buchanan, New York
Utility: Alabama Power Company
Reactor Supplier: Westinghouse Corporation
Capacity: 854 net MWe
Reactor Type: Pressurized water reactor
Date of Operation: March 1981
License Expiration Date: 03/31/2021
Electricity Produced in 1998: 6.3 billion kWh
1998 Average Capacity Factor: 83.8%

KEWAUNEE
Location: Carlton, Wisconsin
Utility: Wisconsin Public Service Corporation
Reactor Supplier: Westinghouse Corporation
Capacity: 498 net MWe
Reactor Type: Pressurized water reactor
Date of Operation: December 1973
License Expiration Date: 12/21/2013
Electricity Produced in 1998: 3.7 billion kWh
1998 Average Capacity Factor: 84.9%

LASALLE 1
Location: Seneca, Illinois
Utility: Commonwealth Edison Company
Reactor Supplier: General Electric Company
Capacity: 1048 net MWe
Reactor Type: Boiling water reactor
Date of Operation: August 1982
License Expiration Date: 05/17/2022
Electricity Produced in 1998: 3.3 billion kWh
1998 Average Capacity Factor: 36.0%

LASALLE 2
Location: Seneca, Illinois
Utility: Commonwealth Edison Company
Reactor Supplier: General Electric Company
Capacity: 1048 net MWe
Reactor Type: Boiling water reactor
Date of Operation: March 1984
License Expiration Date: 12/16/2023
Electricity Produced in 1998: –0.1 billion kWh
1998 Average Capacity Factor: –0.7%[b]

LIMERICK 1
Location: Pottstown, Pennsylvania
Utility: Philadelphia Electric Company (PECO)
Reactor Supplier: General Electric Company
Capacity: 1105 net MWe
Reactor Type: Boiling water reactor
Date of Operation: August 1985
License Expiration Date: 10/26/2024
Electricity Produced in 1998: 7.5 billion kWh
1998 Average Capacity Factor: 77.0%

LIMERICK 2
Location: Pottstown, Pennsylvania
Utility: Philadelphia Electric Company (PECO)
Reactor Supplier: General Electric Company
Capacity: 1115 net MWe
Reactor Type: Boiling water reactor
Date of Operation: August 1989
License Expiration Date: 06/22/2029
Electricity Produced in 1998: 9.3 billion kWh
1998 Average Capacity Factor: 95.1%

MAINE YANKEE (retired)
Location: Wiscasset, Maine
Utility: Maine Yankee Atomic Power Company
Reactor Supplier: Combustion Engineering, Inc.
Capacity: 870 net MWe
Reactor Type: Pressurized water reactor
Date of Operation: June 1973
License Expiration Date: 10/21/2008
Electricity Produced in 1998: 0.0 billion kWh
1998 Average Capacity Factor: 0.0%

MCGUIRE 1
Location: Cowens Ford, North Carolina
Utility: Duke Power Company
Reactor Supplier: Westinghouse Corporation
Capacity: 1129 net MWe
Reactor Type: Pressurized water reactor
Date of Operation: July 1981
License Expiration Date: 06/12/2021

(continues)

TABLE 4.1
(continued)

Electricity Produced in 1998: 8.8 billion kWh
1998 Average Capacity Factor: 89.2%

MCGUIRE 2
Location: Cowens Ford, North Carolina
Utility: Duke Power Company
Reactor Supplier: Westinghouse Corporation
Capacity: 1129 net MWe
Reactor Type: Pressurized water reactor
Date of Operation: May 1983
License Expiration Date: 03/03/2023
Electricity Produced in 1998: 9.9 billion kWh
1998 Average Capacity Factor: 100.4%[a]

MILLSTONE 1 (retired)
Location: Waterford, Connecticut
Utility: Northeast Nuclear Energy Company
Reactor Supplier: General Electric Company
Capacity: 641 net MWe
Reactor Type: Boiling water reactor
Date of Operation: October 1970
License Expiration Date: 10/06/2010
Electricity Produced in 1998: 0.0 billion kWh
1998 Average Capacity Factor: 0.0%

MILLSTONE 2
Location: Waterford, Connecticut
Utility: Northeast Nuclear Energy Company
Reactor Supplier: Combustion Engineering, Inc.
Capacity: 871 net MWe
Reactor Type: Pressurized water reactor
Date of Operation: September 1975
License Expiration Date: 07/31/2015
Electricity Produced in 1998: 0.0 billion kWh
1998 Average Capacity Factor: 0.0%

MILLSTONE 3
Location: Waterford, Connecticut
Utility: Northeast Nuclear Energy Company
Reactor Supplier: Westinghouse Corporation
Capacity: 1120 net MWe
Reactor Type: Pressurized water reactor
Date of Operation: January 1986
License Expiration Date: 11/25/2025
Electricity Produced in 1998: 3.3 billion kWh
1998 Average Capacity Factor: 33.7%

MONTICELLO
Location: Monticello, Minnesota
Utility: Northern States Power Company
Reactor Supplier: General Electric Company
Capacity: 545 net MWe
Reactor Type: Boiling water reactor
Date of Operation: January 1971
License Expiration Date: 09/08/2010
Electricity Produced in 1998: 4.1 billion kWh
1998 Average Capacity Factor: 86.3%

NINE MILE POINT 1
Location: Oswego, New York
Utility: Niagara Mohawk Power Corporation
Reactor Supplier: General Electric Company
Capacity: 618 net MWe
Reactor Type: Boiling water reactor
Date of Operation: August 1969
License Expiration Date: 08/22/2009
Electricity Produced in 1998: 4.8 billion kWh
1998 Average Capacity Factor: 89.5%

NINE MILE POINT 2
Location: Oswego, New York
Utility: Niagara Mohawk Power Corporation
Reactor Supplier: General Electric Company
Capacity: 1136 net MWe
Reactor Type: Boiling water reactor
Date of Operation: July 1987
License Expiration Date: 10/31/2026
Electricity Produced in 1998: 7.3 billion kWh
1998 Average Capacity Factor: 73.4%

NORTH ANNA 1
Location: Mineral, Virginia
Utility: Virginia Electric and Power Company
Reactor Supplier: Westinghouse Corporation
Capacity: 893 net MWe
Reactor Type: Pressurized water reactor
Date of Operation: April 1978
License Expiration Date: 04/01/2018

(continues)

TABLE 4.1
(continued)

Electricity Produced in 1998: 7.2 billion kWh
1998 Average Capacity Factor: 92.3%

NORTH ANNA 2
Location: Mineral, Virginia
Utility: Virginia Electric and Power Company
Reactor Supplier: Westinghouse Corporation
Capacity: 897 net MWe
Reactor Type: Pressurized water reactor
Date of Operation: August 1980
License Expiration Date: 08/21/2020
Electricity Produced in 1998: 7.1 billion kWh
1998 Average Capacity Factor: 90.2%

OCONEE 1
Location: Seneca, South Carolina
Utility: Duke Power Company
Reactor Supplier: Babcock and Wilcox
Capacity: 846 net MWe
Reactor Type: Pressurized water reactor
Date of Operation: February 1973
License Expiration Date: 02/06/2013
Electricity Produced in 1998: 6.0 billion kWh
1998 Average Capacity Factor: 80.8%

OCONEE 2
Location: Seneca, South Carolina
Utility: Duke Power Company
Reactor Supplier: Babcock and Wilcox
Capacity: 846 net MWe
Reactor Type: Pressurized water reactor
Date of Operation: October 1973
License Expiration Date: 10/06/2013
Electricity Produced in 1998: 5.7 billion kWh
1998 Average Capacity Factor: 76.3%

OCONEE 3
Location: Seneca, South Carolina
Utility: Duke Power Company
Reactor Supplier: Babcock and Wilcox
Capacity: 846 net MWe
Reactor Type: Pressurized water reactor
Date of Operation: July 1974
License Expiration Date: 07/19/2014
Electricity Produced in 1998: 5.8 billion kWh
1998 Average Capacity Factor: 78.0%

OYSTER CREEK 1
Location: Forked River, New Jersey
Utility: GPU Nuclear Corporation
Reactor Supplier: General Electric Company
Capacity: 619 net MWe
Reactor Type: Boiling water reactor
Date of Operation: August 1969
License Expiration Date: 12/15/2009
Electricity Produced in 1998: 4.3 billion kWh
1998 Average Capacity Factor: 79.7%

PALISADES
Location: South Haven, Michigan
Utility: Consumers Power Company
Reactor Supplier: Combustion Engineering, Inc.
Capacity: 762 net MWe
Reactor Type: Pressurized water reactor
Date of Operation: October 1972
License Expiration Date: 03/14/2007
Electricity Produced in 1998: 5.4 billion kWh
1998 Average Capacity Factor: 80.4%

PALO VERDE 1
Location: Wintersburg, Arizona
Utility: Arizona Public Service Company
Reactor Supplier: Combustion Engineering, Inc.
Capacity: 1258 net MWe
Reactor Type: Pressurized water reactor
Date of Operation: June 1985
License Expiration Date: 12/31/2024
Electricity Produced in 1998: 9.6 billion kWh
1998 Average Capacity Factor: 87.1%

PALO VERDE 2
Location: Wintersburg, Arizona
Utility: Arizona Public Service Company
Reactor Supplier: Combustion Engineering, Inc.
Capacity: 1258 net MWe
Reactor Type: Pressurized water reactor
Date of Operation: April 1986
License Expiration Date: 12/09/2025

(continues)

TABLE 4.1
(continued)

Electricity Produced in 1998: 11.1 billion kWh
1998 Average Capacity Factor: 100.6%[a]

PALO VERDE 3
Location: Wintersburg, Arizona
Utility: Arizona Public Service Company
Reactor Supplier: Combustion Engineering, Inc.
Capacity: 1262 net MWe
Reactor Type: Pressurized water reactor
Date of Operation: November 1987
License Expiration Date: 03/25/2027
Electricity Produced in 1998: 9.6 billion kWh
1998 Average Capacity Factor: 87.0%

PEACH BOTTOM 2
Location: Lancaster, Pennsylvania
Utility: Philadelphia Electric Company/Public Service Electric and Gas Company
Reactor Supplier: General Electric Company
Capacity: 1093 net MWe
Reactor Type: Boiling water reactor
Date of Operation: December 1973
License Expiration Date: 08/08/2013
Electricity Produced in 1998: 7.6 billion kWh
1998 Average Capacity Factor: 79.9%

PEACH BOTTOM 3
Location: Lancaster, Pennsylvania
Utility: Philadelphia Electric Company/Public Service Electric and Gas Company
Reactor Supplier: General Electric Company
Capacity: 1093 net MWe
Reactor Type: Boiling water reactor
Date of Operation: July 1974
License Expiration Date: 07/02/2014
Electricity Produced in 1998: 8.8 billion kWh
1998 Average Capacity Factor: 92.1%

PERRY 1
Location: North Perry, Ohio
Utility: Cleveland Electric Illuminating Company
Reactor Supplier: General Electric Company
Capacity: 1169 net MWe
Reactor Type: Boiling water reactor
Date of Operation: November 1986
License Expiration Date: 03/18/2026
Electricity Produced in 1998: 10.3 billion kWh
1998 Average Capacity Factor: 100.9%[a]

PILGRIM 1
Location: Plymouth, Massachusetts
Utility: Boston Edison Company
Reactor Supplier: General Electric Company
Capacity: 669 net MWe
Reactor Type: Boiling water reactor
Date of Operation: September 1972
License Expiration Date: 06/08/2012
Electricity Produced in 1998: 5.7 billion kWh
1998 Average Capacity Factor: 97.2%

POINT BEACH 1
Location: Two Creeks, Wisconsin
Utility: Wisconsin Electric Power Company
Reactor Supplier: Westinghouse Corporation
Capacity: 493 net MWe
Reactor Type: Pressurized water reactor
Date of Operation: October 1970
License Expiration Date: 10/05/2010
Electricity Produced in 1998: 2.6 billion kWh
1998 Average Capacity Factor: 59.5%

POINT BEACH 2
Location: Two Creeks, Wisconsin
Utility: Wisconsin Electric Power Company
Reactor Supplier: Westinghouse Corporation
Capacity: 441 net MWe
Reactor Type: Pressurized water reactor
Date of Operation: March 1973
License Expiration Date: 03/08/2013
Electricity Produced in 1998: 3.1 billion kWh
1998 Average Capacity Factor: 80.8%

PRAIRIE ISLAND 1
Location: Red Wing, Minnesota
Utility: Northern States Power Company
Reactor Supplier: Westinghouse Corporation
Capacity: 514 net MWe
Reactor Type: Pressurized water reactor
Date of Operation: April 1974
License Expiration Date: 08/09/2013

(continues)

TABLE 4.1
(continued)

Electricity Produced in 1998: 4.2 billion kWh
1998 Average Capacity Factor: 93.2%

PRAIRIE ISLAND 2
Location: Red Wing, Minnesota
Utility: Northern States Power Company
Reactor Supplier: Westinghouse Corporation
Capacity: 513 net MWe
Reactor Type: Pressurized water reactor
Date of Operation: October 1974
License Expiration Date: 10/29/2014
Electricity Produced in 1998: 3.3 billion kWh
1998 Average Capacity Factor: 74.0%

QUAD CITIES 1 and 2
Location: Cordova, Illinois
Utility: Commonwealth Edison Company
Reactor Supplier: General Electric Company
Capacity: 769 net MWe
Reactor Type: Boiling water reactors
Date of Operation: December 1972
License Expiration Date: 12/14/2012
Electricity Produced in 1998: Quad Cities 1–3.1 billion kWh, Quad Cities 2–3.8 billion kWh
1998 Average Capacity Factor: Quad Cities 1–46.5%, Quad Cities 2–56.7%

RIVER BEND 1
Location: St. Francisville, Louisiana
Utility: Gulf States Utilities Company
Reactor Supplier: General Electric Company
Capacity: 936 net MWe
Reactor Type: Boiling water reactor
Date of Operation: November 1985
License Expiration Date: 08/29/2025
Electricity Produced in 1998: 7.8 billion kWh
1998 Average Capacity Factor: 95.4%

ROBERT E. GINNA
Location: Rochester, New York
Utility: Rochester Gas and Electric Corporation
Reactor Supplier: Westinghouse Corporation
Capacity: 485 net MWe
Reactor Type: Pressurized water reactor
Date of Operation: September 1969
License Expiration Date: 09/18/2009
Electricity Produced in 1998: 4.1 billion kWh
1998 Average Capacity Factor: 96.8%

SALEM 1
Location: Salem, New Jersey
Utility: Public Service Electric and Gas Company
Reactor Supplier: Westinghouse Corporation
Capacity: 1106 net MWe
Reactor Type: Pressurized water reactor
Date of Operation: December 1976
License Expiration Date: 08/13/2016
Electricity Produced in 1998: 6.5 billion kWh
1998 Average Capacity Factor: 66.8%

SALEM 2
Location: Salem, New Jersey
Utility: Public Service Electric and Gas Company
Reactor Supplier: Westinghouse Corporation
Capacity: 1106 net MWe
Reactor Type: Pressurized water reactor
Date of Operation: May 1981
License Expiration Date: 04/18/2020
Electricity Produced in 1998: 7.6 billion kWh
1998 Average Capacity Factor: 78.6%

SAN ONOFRE 2
Location: San Clemente, California
Utility: Southern California Edison Company
Reactor Supplier: Combustion Engineering, Inc.
Capacity: 1070 net MWe
Reactor Type: Pressurized water reactor
Date of Operation: September 1982
License Expiration Date: 10/18/2013
Electricity Produced in 1998: 8.4 billion kWh
1998 Average Capacity Factor: 89.9%

SAN ONOFRE 3
Location: San Clemente, California
Utility: Southern California Edison Company
Reactor Supplier: Combustion Engineering, Inc.
Capacity: 1080 net MWe
Reactor Type: Pressurized water reactor
Date of Operation: September 1983
License Expiration Date: 10/18/2013

(continues)

TABLE 4.1
(continued)

Electricity Produced in 1998: 9.1 billion kWh
1998 Average Capacity Factor: 95.7%

SEABROOK 1
Location: Seabrook, New Hampshire
Utility: Public Service Company of New Hampshire
Reactor Supplier: Westinghouse Corporation
Capacity: 1162 net MWe
Reactor Type: Pressurized water reactor
Date of Operation: March 1990
License Expiration Date: 10/17/2026
Electricity Produced in 1998: 8.4 billion kWh
1998 Average Capacity Factor: 82.4%

SEQUOYAH 1
Location: Daisy, Tennessee
Utility: Tennessee Valley Authority
Reactor Supplier: Westinghouse Corporation
Capacity: 1119 net MWe
Reactor Type: Pressurized water reactor
Date of Operation: September 1980
License Expiration Date: 09/17/2020
Electricity Produced in 1998: 8.9 billion kWh
1998 Average Capacity Factor: 90.9%

SEQUOYAH 2
Location: Daisy, Tennessee
Utility: Tennessee Valley Authority
Reactor Supplier: Westinghouse Corporation
Capacity: 1119 net MWe
Reactor Type: Pressurized water reactor
Date of Operation: September 1981
License Expiration Date: 09/15/2021
Electricity Produced in 1998: 9.8 billion kWh
1998 Average Capacity Factor: 100.0%

SHEARON HARRIS 1
Location: New Hill, North Carolina
Utility: Carolina Power and Light Company
Reactor Supplier: Westinghouse Corporation
Capacity: 860 net MWe
Reactor Type: Pressurized water reactor
Date of Operation: January 1987
License Expiration Date: 10/24/2026

Electricity Produced in 1998: 6.7 billion kWh
1998 Average Capacity Factor: 89.1%

SOUTH TEXAS 1
Location: Bay City, Texas
Utility: Houston Lighting and Power Company
Reactor Supplier: Westinghouse Corporation
Capacity: 1250 net MWe
Reactor Type: Pressurized water reactor
Date of Operation: March 1988
License Expiration Date: 08/20/2027
Electricity Produced in 1998: 10.9 billion kWh
1998 Average Capacity Factor: 99.2%

SOUTH TEXAS 2
Location: Bay City, Texas
Utility: Houston Lighting and Power Company
Reactor Supplier: Westinghouse Corporation
Capacity: 1250 net MWe
Reactor Type: Pressurized water reactor
Date of Operation: March 1989
License Expiration Date: 12/15/2028
Electricity Produced in 1998: 10.0 billion kWh
1998 Average Capacity Factor: 91.1%

ST. LUCIE 1
Location: Ft. Pierce, Florida
Utility: Florida Power and Light Company
Reactor Supplier: Combustion Engineering, Inc.
Capacity: 839 net MWe
Reactor Type: Pressurized water reactor
Date of Operation: March 1976
License Expiration Date: 03/01/2016
Electricity Produced in 1998: 7.0 billion kWh
1998 Average Capacity Factor: 94.9%

ST. LUCIE 2
Location: Ft. Pierce, Florida
Utility: Florida Power and Light Company
Reactor Supplier: Combustion Engineering, Inc.
Capacity: 839 net MWe
Reactor Type: Pressurized water reactor
Date of Operation: June 1983
License Expiration Date: 04/06/2023

(continues)

TABLE 4.1
(continued)

Electricity Produced in 1998: 6.7 billion kWh
1998 Average Capacity Factor: 90.8%

SUMMER 1
Location: Jenkinsville, South Carolina
Utility: South Carolina Electric and Gas Company
Reactor Supplier: Westinghouse Corporation
Capacity: 942 net MWe
Reactor Type: Pressurized water reactor
Date of Operation: November 1982
License Expiration Date: 08/06/2022
Electricity Produced in 1998: 8.2 billion kWh
1998 Average Capacity Factor: 99.0%

SURRY 1
Location: Surry, Virginia
Utility: Virginia Electric and Power Company
Reactor Supplier: Westinghouse Corporation
Capacity: 801 net MWe
Reactor Type: Pressurized water reactor
Date of Operation: May 1972
License Expiration Date: 05/25/2012
Electricity Produced in 1998: 5.8 billion kWh
1998 Average Capacity Factor: 82.0%

SURRY 2
Location: Surry, Virginia
Utility: Virginia Electric and Power Company
Reactor Supplier: Westinghouse Corporation
Capacity: 801 net MWe
Reactor Type: Pressurized water reactor
Date of Operation: January 1973
License Expiration Date: 01/29/2013
Electricity Produced in 1998: 7.2 billion kWh
1998 Average Capacity Factor: 102.3%[a]

SUSQUEHANNA 1
Location: Berwick, Pennsylvania
Utility: Pennsylvania Power and Light Company (PP&L)
Reactor Supplier: General Electric Company
Capacity: 1090 net MWe
Reactor Type: Boiling water reactor
Date of Operation: November 1982
License Expiration Date: 07/17/2022
Electricity Produced in 1998: 7.6 billion kWh
1998 Average Capacity Factor: 79.4%

SUSQUEHANNA 2
Location: Berwick, Pennsylvania
Utility: Pennsylvania Power and Light Company (PP&L)
Reactor Supplier: General Electric Company
Capacity: 1094 net MWe
Reactor Type: Boiling water reactor
Date of Operation: June 1984
License Expiration Date: 03/23/2024
Electricity Produced in 1998: 8.8 billion kWh
1998 Average Capacity Factor: 92.0%

THREE MILE ISLAND 1
Location: Middletown, Pennsylvania
Utility: GPU Nuclear Corporation
Reactor Supplier: Babcock and Wilcox
Capacity: 786 net MWe
Reactor Type: Pressurized water reactor
Date of Operation: April 1974
License Expiration Date: 04/19/2014
Electricity Produced in 1998: 7.1 billion kWh
1998 Average Capacity Factor: 102.5%[a]

TURKEY POINT 3
Location: Florida City, Florida
Utility: Florida Power and Light Company
Reactor Supplier: Westinghouse Corporation
Capacity: 693 net MWe
Reactor Type: Pressurized water reactor
Date of Operation: July 1972
License Expiration Date: 07/19/2012
Electricity Produced in 1998: 5.4 billion kWh
1998 Average Capacity Factor: 89.1%

TURKEY POINT 4
Location: Florida City, Florida
Utility: Florida Power and Light Company
Reactor Supplier: Westinghouse Corporation
Capacity: 693 net MWe
Reactor Type: Pressurized water reactor
Date of Operation: April 1973

(continues)

TABLE 4.1
(continued)

License Expiration Date: 04/10/2013
Electricity Produced in 1998: 6.2 billion kWh
1998 Average Capacity Factor: 101.9%[a]

VERMONT YANKEE 1
Location: Vernon, Vermont
Utility: Vermont Yankee Nuclear Power Corporation
Reactor Supplier: General Electric Company
Capacity: 496 net MWe
Reactor Type: Boiling water reactor
Date of Operation: February 1973
License Expiration Date: 03/21/2012
Electricity Produced in 1998: 3.4 billion kWh
1998 Average Capacity Factor: 77.3%

VOGTLE 1
Location: Waynesboro, Georgia
Utility: Georgia Power Company
Reactor Supplier: Westinghouse Corporation
Capacity: 1164 net MWe
Reactor Type: Pressurized water reactor
Date of Operation: March 1987
License Expiration Date: 01/16/2027
Electricity Produced in 1998: 10.2 billion kWh
1998 Average Capacity Factor: 100.2%[a]

VOGTLE 2
Location: Waynesboro, Georgia
Utility: Georgia Power Company
Reactor Supplier: Westinghouse Corporation
Capacity: 1169 net MWe
Reactor Type: Pressurized water reactor
Date of Operation: March 1989
License Expiration Date: 02/09/2029
Electricity Produced in 1998: 8.4 billion kWh
1998 Average Capacity Factor: 81.9%

WASHINGTON NUCLEAR 2
Location: Richland, Washington
Utility: Washington Public Power Supply System
Reactor Supplier: General Electric Company
Capacity: 1170 net MWe
Reactor Type: Boiling water reactor
Date of Operation: April 1984
License Expiration Date: 12/20/2023
Electricity Produced in 1998: 6.9 billion kWh
1998 Average Capacity Factor: 67.5%

WATERFORD 3
Location: Taft, Louisiana
Utility: Louisiana Power and Light Company
Reactor Supplier: Combustion Engineering, Inc.
Capacity: 1075 net MWe
Reactor Type: Pressurized water reactor
Date of Operation: March 1985
License Expiration Date: 12/18/2024
Electricity Produced in 1998: 8.6 billion kWh
1998 Average Capacity Factor: 91.3%

WATTS BAR 1
Location: Spring City, Tennessee
Utility: Tennessee Valley Authority
Capacity: 1119 net MWe
Reactor Type: Pressurized water reactor
Date of Operation: February 1996
License Expiration Date: 11/09/2035
Electricity Produced in 1998: 9.7 billion kWh
1998 Average Capacity Factor: 98.8%

WOLF CREEK
Location: Burlington, Kansas
Utility: Wolf Creek Nuclear Operating Corporation
Reactor Supplier: Westinghouse Corporation
Capacity: 1163 net MWe
Reactor Type: Pressurized water reactor
Date of Operation: June 1985
License Expiration Date: 03/11/2025
Electricity Produced in 1998: 10.4 billion kWh
1998 Average Capacity Factor: 102.2%[a]

ZION 1 (retired)
Location: Zion, Illinois
Utility: Commonwealth Edison Company
Reactor Supplier: Westinghouse Corporation
Capacity: 1040 net MWe
Reactor Type: Pressurized water reactor
Date of Operation: October 1973
License Expiration Date: 04/06/2013
Electricity Produced in 1998: 0.0 billion kWh
1998 Average Capacity Factor: 0.0%

(continues)

TABLE 4.1
(continued)

ZION 2 (retired)	Reactor Type: Pressurized water reactor
Location: Zion, Illinois	Date of Operation: November 1973
Utility: Commonwealth Edison Company	License Expiration Date: 11/14/2013
Reactor Supplier: Westinghouse Corporation	Electricity Produced in 1998: 0.0 billion kWh
Capacity: 1040 net MWe	1998 Average Capacity Factor: 0.0%

Notes: Operating reactors are those licensed by the Nuclear Regulatory Commission.
"Date of operation" is the date the unit received either a provisional or a full-power license.
A more detailed version of this information, including site descriptions, is available for download from the United States Department of Energy at http: //www.eia.doe.gov/cneaf/nuclear/page/at_a_glance/reactors/states.html.
For additional technical data about plant designs and characteristics, see the Nuclear Regulatory Agency's Plant Information Books, available on-line at http: //www.nrc.gov/AEOD/pib/pib.html.
[a] A capacity factor over 100% is possible when a reactor operates every day of the year or when a reactor generates more electricity than its designated power rating.
[b] A negative generation implies that the reactor was not operating and needed to purchase power.
Sources: Location, Utility, Reactor Supplier, Reactor Type, Date of Operation—Nuclear Power Generation and Fuel Cycle Report 1997 (September 1997); Capacity—DOE/EIA-0095(98) Inventory of Power Plants in the United States As of January 1, 1998 (December 1998); License Expiration Date—NRC Information Digest 1997; Electricity Produced—Energy Information Administration, Form EIA-759, "Monthly Power Plant Report"; 1998 Average Capacity Factor—Computed from NRC data; see Nuclear Power Generation and Fuel Cycle Report 1997 for comprehensive information on the international nuclear industry.

World Nuclear Reactors

Figure 4.5 and Tables 4.2 and 4.3 give basic information about the use of nuclear power around the world.

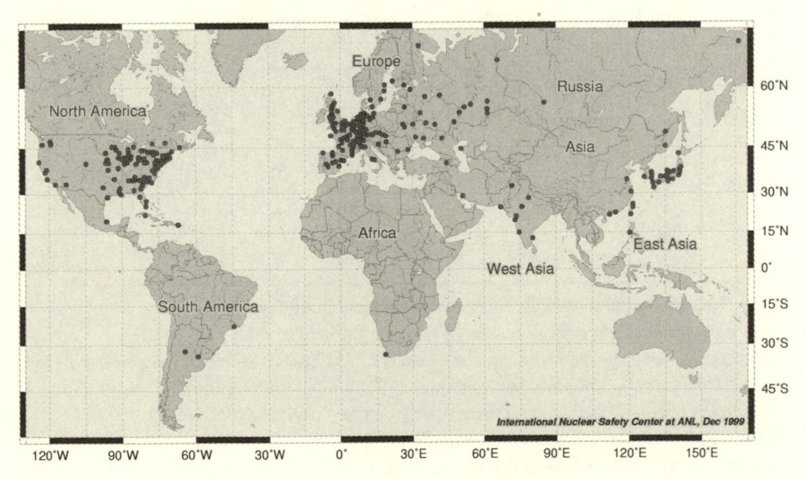

Figure 4.5 World reactors (Argonne National Laboratory)

TABLE 4.2
Statistics on World Nuclear Reactors (showing number of reactors operating in each category and total net output in megawatts)

	Operating		Not Operating		Under Construction		Construction Suspended	
Country or Area	No.	MWe net	No.	MWe net	No.	MWe net	No.	MWe net
Argentina	2	945	—	—	1	692	—	—
Armenia	1	376	1	376	—	—	—	—
Belgium	7	5,713	—	—	—	—	—	—
Brazil	1	626	—	—	1	1,245	—	—
Bulgaria	6	3,526	—	—	—	—	—	—
Canada	14	10,298	8	5,136	—	—	—	—
China	3	2,100	—	—	5	4,000	—	—
Cuba	—	—	—	—	—	—	2	816
Czech Republic	4	1,648	—	—	2	1,824	—	—
Finland	4	2,550	—	—	—	—	—	—
France	58	61,723	1	1,450	—	—	—	—
Germany	19	21,044	1	1,219	—	—	—	—
Hungary	4	1,720	—	—	—	—	—	—
India	11	1,997	1	200	4	1,300	—	—
Iran	—	—	—	—	1	950	1	950
Japan	52	43,249	1	246	1	796	—	—
Kazakhstan	—	—	—	—	—	—	—	—
Korea, S.	15	12,030	380	—	—	5	4	800
Lithuania	2	2,500	—	—	—	—	—	—
Mexico	2	1,308	—	—	—	—	—	—
Netherlands	1	449	—	—	—	—	—	—
Pakistan	1	125	—	—	1	300	—	—
Romania	1	630	—	—	—	—	4	2,520
Russia	29	19,843	—	—	3	2,825	7	6,628
Slovakia	6	4,430	—	—	—	—	2	810
Slovenia	1	620	—	—	—	—	—	—
South Africa	2	1,840	—	—	—	—	—	—
Spain	9	7,394	—	—	—	—	—	—
Sweden	11	9,325	—	—	—	—	—	—
Switzerland	5	3,122	—	—	—	—	—	—
Taiwan	6	4,884	—	—	—	—	—	—
UK	35	12,996	—	—	—	—	—	—
Ukraine	14	12,153	—	—	2	1,906	3	2,859
USA	104	96,977	2	1,935	—	—	6	7,293
Total	430	346,141	15	10,562	36	20,638	25	21,876

Notes: France: Phenix FBR is included in the number of "operating" reactors.
China: Lianyungang—1 now under construction.
India: Kaiga—2 declared in commercial operation December 25, 1999, but not expected to reach full capacity until March 2000. Rajasthan—3 reached criticality December 25, 1999.
South Korea: Wolsong—4 declared in commercial operation October 1, 1999.
Slovakia: Mochovce—2 connected to grid December 1999.
Sweden: Barseback—1 closed down on November 30, 1999.
Source: The Uranium Institute

TABLE 4.3
Percentage of Electricity Generated from Nuclear Power in 1998

Country	Percent	Country	Percent
Argentina	10.04	Lithuania	77.21
Armenia	24.69	Mexico	5.41
Belgium	55.16	Netherlands	4.13
Brazil	1.08	Pakistan	0.65
Bulgaria	41.50	Romania	10.35
Canada	12.44	Russia	13.08
China	1.16	Slovakia	43.80
Czech Republic	20.50	Slovenia	38.33
Finland	27.44	South Africa	7.25
France	75.77	Spain	31.66
Germany	28.29	Sweden	45.75
Hungary	35.62	Switzerland	41.07
India	2.51	Taiwan	24.77
Japan	35.86	UK	27.09
Kazakhstan	0.18	Ukraine	45.42
Korea, S.	41.39	USA	18.69

Source: The Uranium Institute

The Fuel Cycle and Nuclear Waste

The following documents and illustrations provide an overview of the extent of the problem of nuclear waste and the issues involved in waste management. See Figure 4.6 for a diagram of the nuclear fuel cycle.

The Nuclear Fuel Cycle

The nuclear fuel cycle is the series of industrial processes which facilitate the production of electricity from uranium in nuclear power reactors. Uranium is a relatively common element that is found throughout the world. It is mined in a number of countries and must be processed before it can be used as fuel for a nuclear reactor. Electricity is created by using the heat generated in a nuclear reactor to produce steam and drive a turbine connected to a generator. Fuel removed from a reactor, after it has reached the end of its useful life, can be reprocessed to produce new fuel.

The various activities associated with the production of electricity from nuclear reactions are referred to collectively as the nuclear fuel cycle. The nuclear fuel cycle starts with the mining of uranium and ends with the disposal of nuclear waste. With the introduction of reprocessing as an option for nuclear fuel, the stages can now form a true cycle.

Uranium

Uranium is a slightly radioactive metal that occurs throughout the earth's crust. It is about 500 times more abundant than gold and about as common as tin. It is present in most rocks and soils as well as in many rivers and in sea water. It is, for example, found in concentrations of about four parts per million (ppm) in granite, which makes up 60% of the earth's crust. In fertilisers, uranium concentration can be as high as 400 ppm (0.04%), and some coal deposits contain uranium at concentrations greater than 100 ppm (0.01%).

There are a number of areas around the world where the concentration of uranium in the ground is sufficiently high that extraction for use as nuclear fuel is economically feasible.

Uranium Mining

Both underground and open pit mining techniques are used to recover uranium. In general, open pit mining is used where deposits are close to the surface and underground mining is used for deep deposits, typically greater than 120 m deep. Open pit mines require large openings on the surface, larger than the size of the ore deposit, since the walls of the pit must be sloped to guard against collapse. As a result, the quantity of material that must be removed in order to access the ore is large. Underground mines have relatively small openings to the surface and the quantity of material that must be removed to access the ore is considerably less than in the case of an open pit mine.

The decision as to which mining method to use for a particular deposit is governed by safety and economic considerations. In the case of underground uranium mines special precautions, consisting primarily of increased ventilation, are required to protect against airborne radiation exposure.

Uranium Milling

Milling, which is generally carried out close to a uranium mine, extracts the uranium from the ore. Most mining facilities include a mill, although where mines are close together, one mill may process the ore from several mines. Milling produces a uranium concentrate that has a smaller volume than the ore, and hence is less expensive to ship.

The uranium concentrate shipped from a mill, commonly referred to as 'yellowcake', generally contains more than 60% uranium. The original uranium ore contains typically between 0.1 and 1% uranium.

In a mill, uranium is extracted from the ore by a combination of one or more chemical processes. The most common process is leaching, in which either a strong acid or a strong alkaline solution is used to dissolve the uranium, which is then precipitated from solution as a concentrate.

Some uranium is produced by direct chemical extraction rather than mining. One method by which this is achieved is in situ leaching of uranium deposits. Where this method is employed, a chemical solution in which uranium will dissolve is introduced into the ground at or close to a deposit. Ground water is then extracted from the area using a well which has been located such that the uranium solvent is drawn through the ore body. This approach is only appropriate where geological conditions are such that ground water contamination can be avoided.

The chemical solution dissolves the uranium and carries it to the well. The uranium can then be precipitated from the resulting solution in a manner similar to that which is used in a mill.

Conversion

The product of a uranium mill is not directly usable as a fuel for a nuclear reactor. Additional processing, generally referred to as conversion, is required.

At a conversion facility, uranium is converted to either uranium dioxide, which can be used as the fuel for those types of reactors that do not require enriched uranium, or into uranium hexafluoride, which can be enriched to produce fuel for the majority of types of reactors.

Enrichment

Natural uranium consists, primarily, of a mixture of two isotopes (forms) of uranium. Only 0.7% of natural uranium is capable of undergoing fission, the process by which energy is produced in a nuclear reactor. The fissionable isotope of uranium is uranium 235 (U-235). The remainder is uranium 238 (U-238).

In the most common types of nuclear reactors, a higher than natural concentration of U-235 is required. The enrichment process produces this higher concentration, typically between 3.5% and 4.5% U-235, by removing a large part of the U-238 (80% for enrichment to 3.5%).

There are two enrichment processes in large scale commercial use, each of which uses uranium hexafluoride as feed: gaseous diffusion and gas centrifugation. The product of this stage of the nuclear fuel cycle is enriched uranium hexafluoride, which is reconverted to produce enriched uranium oxide.

Fuel Fabrication

Reactor fuel is generally in the form of ceramic pellets. These are formed from pressed uranium oxide which is sintered at a high temperature (over 1400 °C). The pellets are then encased in metal tubes, which are arranged into a fuel assembly ready for introduction into a reactor.

The dimensions of the fuel pellets and other components of the fuel assembly are precisely controlled to ensure consistency in the characteristics of fuel bundles.

Power Generation

Inside a nuclear reactor the nuclei of U-235 atoms split (fission) and, in the process, release energy. This energy is used to heat water and turn it into steam. The steam is used to drive a turbine connected to a generator which produces electricity. The fissioning of uranium is used as a source of heat in a nuclear power station in the same way that the burning of coal, gas or oil is used as a source of heat in a thermal power plant.

Spent Fuel

With time, the concentration of fission fragments in a fuel bundle will increase to the point where it is no longer practical to continue to use the fuel. At this point the 'spent fuel' is removed from the reactor. The amount of energy that is produced from a fuel bundle varies with the type of reactor and the policy of the reactor operator.

Typically, more than 40 million kilowatt-hours of electricity are produced from one tonne of natural uranium. The production of this amount of electrical power from fossil fuels would require the burning of over 16,000 tonnes of black coal or 80,000 barrels of oil.

Spent Fuel Storage

When removed from a reactor, a fuel bundle will be emitting both radiation, primarily from the fission fragments, and heat. Spent fuel is unloaded into a storage facility immediately adjacent to the reactor to allow the radiation levels and the quantity of heat being released to decrease.

These facilities are large pools of water; the water acts as both a shield against the radiation and an absorber of the heat released. Spent fuel is generally held in such pools for a minimum of about five months.

While much of the spent fuel is held at reactor sites beyond the initial storage period, some of it is transferred to interim storage facilities. Ultimately, spent fuel must either be reprocessed or sent for permanent disposal.

Reprocessing

Spent fuel is about 95% U-238 but it also contains U-235 that has not fissioned, plutonium and fission products, which are highly radioactive. In a reprocessing facility the spent fuel is separated into its three components: uranium, plutonium and waste, containing fission products. Reprocessing facilitates recycling and produces a significantly reduced volume of waste.

Uranium and Plutonium Recycling

The uranium from reprocessing, which typically contains a slightly higher concentration of U-235 than occurs in nature, can be reused as fuel after conversion and enrichment, if necessary. The plutonium can be made into mixed oxide (mox) fuel, in which uranium and plutonium oxides are combined.

In reactors that use mox fuel, plutonium substitutes for U-235 as the material that fissions and produces heat for steam production and neutrons to sustain a chain reaction.

Spent Fuel Disposal

At the present time, there are no disposal facilities (as opposed to storage facilities) in operation in which spent fuel, not destined for reprocessing, and the waste from reprocessing can be placed. Although technical issues related to disposal have been addressed, there is currently no pressing need to establish such facilities, as the total volume of such wastes is relatively small. Further, there is a reluctance to dispose of spent fuel because it represents a resource, which could be reprocessed at a later date to allow recycling of the uranium and plutonium.

A number of countries are carrying out studies to determine the optimum approach to the disposal of spent fuel and waste from reprocessing. The most commonly favoured method for disposal being contemplated is placement into deep geological formations.

Wastes

Wastes from the nuclear fuel cycle are categorised as high-, medium- or low-level waste by the amount of radiation that they emit. These wastes come from a number of sources and include: essentially non-radioactive waste resulting from mining low-level waste produced at all stages of the fuel cycle [and] intermediate-level waste produced during reactor operation and by reprocessing high-level waste, which is spent fuel and waste containing fission products from reprocessing. Milling wastes contain long-lived radioactive materials in low concentrations and toxic materials such as heavy metals; however, the total quantity of radioactive elements in milling waste is less than in the original ore. These wastes require safe management in order to isolate them from the environment or to ensure that releases are limited to reduce environmental impact. Commonly, shallow burial in engineered facilities is used to dispose of milling wastes.

The enrichment process leads to the production of 'depleted' uranium. This is uranium in which the concentration of U-235 is significantly less than the 0.7% found in nature. Small quantities of this material, which is primarily U-238, are used in applications where high density material is required, including radiation shielding, and some is used in the production of mox. While U-238 is not fissionable it is a low specific activity radioactive material and some precautions must, therefore, be taken in its storage or disposal.

[Note]: Uranium concentrations are sometimes expressed in terms of U3O8 content (U3O8 is an approximation of the chemical composition of typical naturally occurring oxides of uranium). A product that is said to be 60% U3O8 contains 51% uranium metal.

Source: Fact sheet "The Nuclear Fuel Cycle," courtesy of the Uranium Institute

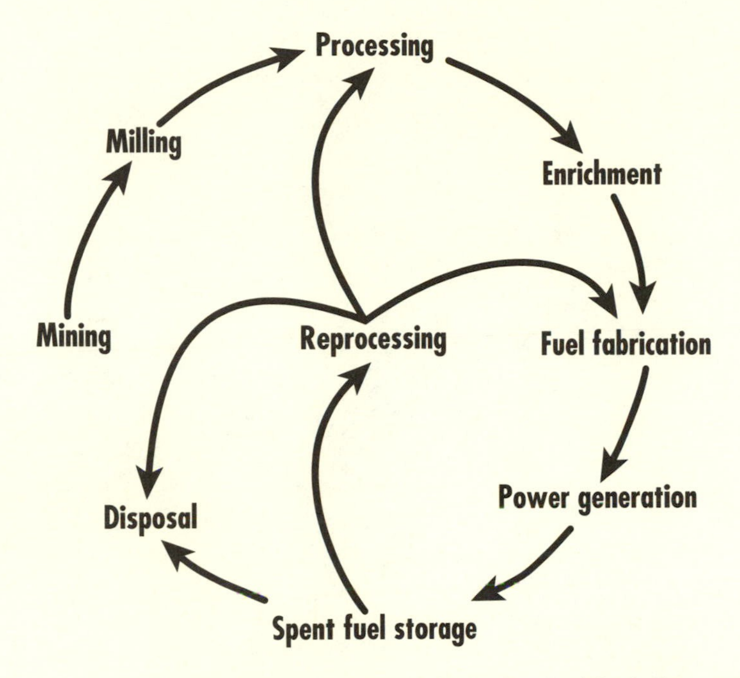

Figure 4.6 Nuclear fuel cycle (fact sheet "The Nuclear Fuel Cycle," courtesy of the Uranium Institute)

Locations of Spent Fuel Storage

Figure 4.7 shows the current location of temporary storage facilities for nuclear waste. The intention is to eventually transport all high-level waste to the facility being developed at Yucca Mountain, Nevada.

The Yucca Mountain Proposal

The following report serves as both an assessment and a sort of "prospectus" for the controversial Yucca Valley nuclear waste storage facility.

Yucca Mountain Nuclear Waste Depository

. . .

The Nuclear Waste Problem

Countries worldwide have accumulated high-level radioactive waste by using nuclear materials to produce electricity, to power

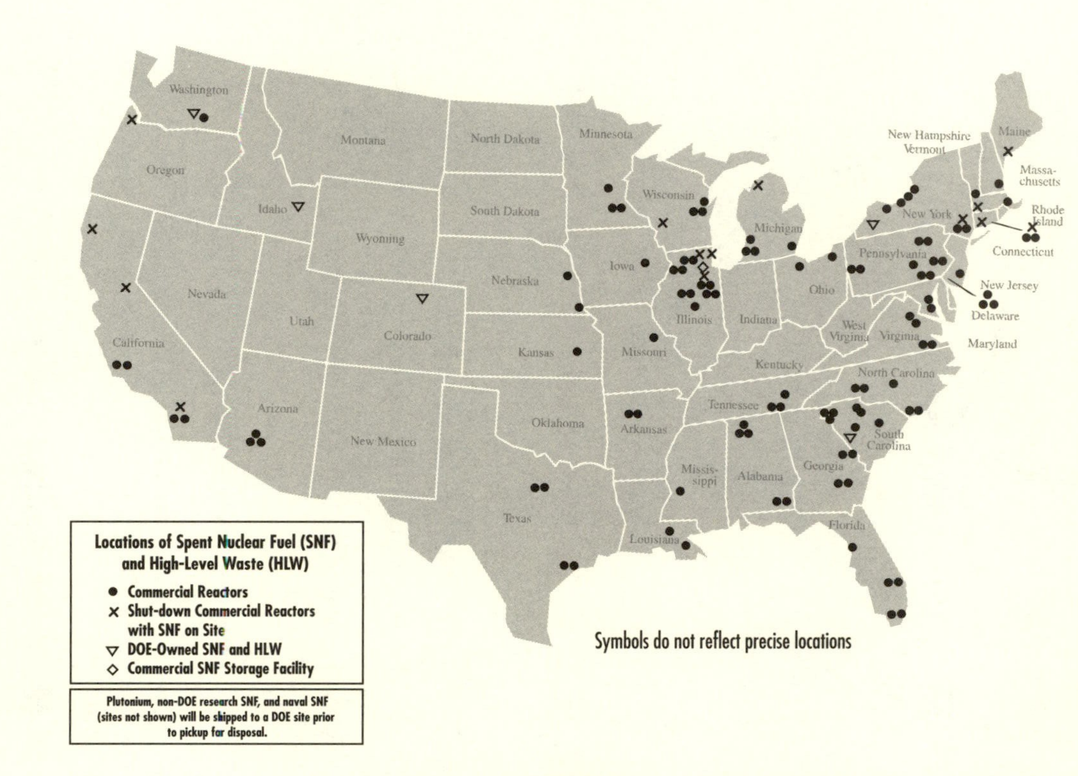

Figure 4.7 Locations of spent fuel storage (DOE/NRC Yucca Mountain Viability Assessment)

naval vessels, and to make nuclear weapons. Some elements of this waste are hazardous for a few years to several hundred years; some elements are hazardous for many thousands of years. This waste must be safely contained until it no longer poses a significant risk to human health and the environment.[2]

Commercial spent nuclear fuel. As of December 1998, the United States had accumulated 38,500 metric tons of used or "spent" nuclear fuel from commercial nuclear power plants; this amount could more than double by the year 2035 if all currently operating plants complete their initial 40-year license period. The spent fuel is now stored in 33 states at 72 power plant sites and one commercial storage site and is likely to remain where it is until a disposal or central storage facility is constructed. When a power plant ceases operations, the spent nuclear fuel and other radioactive materials must be removed before the plant can be fully decommissioned and the site used for other purposes.

DOE spent nuclear fuel. By 2035, the United States will have accumulated approximately 2,500 metric tons of spent nuclear fuel from reactors that produce materials for nuclear weapons, from research reactors, and from reactors on the Navy's nuclear-powered ships and submarines. The majority of DOE spent nuclear fuel is currently stored at three major sites in Idaho, South Carolina, and Washington. Under a negotiated settlement agreement between the State of Idaho, the Navy, and DOE, all spent fuel must be removed from Idaho by the year 2035.[3]

High-level radioactive waste. The production of nuclear weapons has left a legacy of high-level radioactive waste that was created when spent nuclear fuel was treated chemically to separate uranium and plutonium. The remaining high-level waste is in liquid and solid forms; 100 million gallons are stored in underground tanks in Washington, South Carolina, Idaho, and New York.[4] Under agreements between DOE and the states where the waste is stored, this high-level waste will continue to be solidified and placed in about 20,000 canisters for future disposal in a permanent geologic repository.

Surplus plutonium and other nuclear weapons materials. The end of the Cold War has brought the problem of cleaning up and closing weapons plants that are no longer needed and disposing of surplus plutonium and other nuclear materials associated with weapons production. These radioactive materials must be disposed of in a secure facility that will not only keep the waste away from people but will also keep people away from the weapons-usable material for thousands of years. Ensuring national security and preventing the proliferation of nuclear weapons depends on developing a permanent, safe, and secure disposal facility for surplus plutonium and other weapons materials.

Total inventory. At present, spent nuclear fuel and high-level radioactive waste are temporarily stored at 78 locations in 35 states. . . . Some of these storage sites are close to population centers and are located near rivers, lakes, and seacoasts. The stored materials, if left where they are indefinitely, could become a hazard to nearby populations and the environment. These nuclear materials require safe and permanent disposal. . . .

The Law and the Regulations

The Nuclear Waste Policy Act of 1982 (NWPA) directed DOE to develop a system for the safe and final disposal of spent nuclear fuel and high-level radioactive waste.

The NWPA set an ambitious schedule for DOE to site two geologic repositories and required DOE to contract with utilities to begin disposal in the first one by January 31, 1998. The DOE formally identified nine potentially acceptable sites across the nation and later narrowed the list to three sites: Deaf Smith County, Texas; Hanford, Washington; and Yucca Mountain, Nevada. In 1987, Congress directed DOE to study only one of the sites—the one at Yucca Mountain—to decide whether it is suitable for a repository. This legislation, known as the Nuclear Waste Policy Amendments Act of 1987,[10] also established the Nuclear Waste Technical Review Board, composed of experts appointed by the President to review the DOE program.

The NWPA reaffirms the Federal Government's responsibility for developing repositories for the permanent disposal of spent nuclear fuel and high-level radioactive waste. It also affirms the responsibility of the generators of the waste—the nuclear utilities and the federal defense nuclear program—to pay for that effort. The NWPA requires utilities with nuclear power plants to pay a fee to fund the disposal program. The Federal Government bears the costs of disposing of defense waste.

The NWPA also assigns distinct roles to the Environmental Protection Agency (EPA) and the Nuclear Regulatory Commission (NRC). The EPA is required to establish standards for protection of the general environment from releases of radioactive material from a repository. The NRC is responsible for establishing technical requirements and criteria, consistent with EPA standards, for approving or disapproving applications to construct, operate, and eventually close a repository. In 1981 and 1983, NRC issued regulations for a geologic repository in anticipation of EPA standards.[11]

Subsequently, the Energy Policy Act of 1992[12] modified the process for setting environmental standards for a repository at Yucca Mountain. The Act directed the National Academy of Sciences (NAS) to provide findings and recommendations on these standards and directed EPA to issue standards for the Yucca Mountain site based on and consistent with the NAS findings and recommendations. The Act directed NRC to revise its regulations as necessary to be consistent with the EPA standards, once issued. The NAS published its report in 1995.[13] The EPA is currently developing its standards.

How Geologic Disposal Would Work

The basic concept of geologic disposal is to place carefully prepared and packaged waste in excavated tunnels in geologic formations such as salt, hard rock, or clay. The concept relies on a series of barriers, natural and engineered, to contain the waste for thousands of years and to minimize the amount of radioactive material that may eventually be transported from a repository and reach the human environment.

Water is the primary means by which radionuclides could reach the human environment. Therefore, the primary functions of the barriers are to keep water away from the waste as long as possible, to limit the amount of water that finally does contact the waste, to slow the release of radionuclides from the waste, and to reduce the concentrations of radionuclides in groundwater.

All countries pursuing geologic disposal are taking the multibarrier approach, though they differ in the barriers they emphasize. The German disposal concept, for example, relies heavily on the geologic barrier, the rock salt formation at the prospective disposal site. The Swedish method, on the other hand, relies heavily on thick copper waste packages to contain waste.

The U.S. approach, as recommended in the 1979 Report to the President by the Interagency Review Group on Nuclear Waste Management,[14] is to design a repository in which the natural and engineered barriers work as a system, so that some barriers will continue to work even if others fail, and so that none of the barriers is likely to fail for the same reason or at the same time. This design strategy is called defense in depth. The barriers include the chemical and physical forms of the waste, the waste packages and other engineered barriers, and the natural characteristics of Yucca Mountain.

Why Yucca Mountain?

Yucca Mountain is remote from population centers. Located about 100 miles northwest of Las Vegas, Nevada, Yucca Mountain is on the edge of the nation's nuclear weapons test site, where more than 900 nuclear tests have been conducted. This unpopulated land is owned by the Federal Government.

Yucca Mountain provides a stable geologic environment. A flat-topped ridge running six miles from north to south, Yucca Mountain has changed little over the last million years. Based upon what is known about the site, disruption of a repository at Yucca Mountain by volcanoes, earthquakes, erosion, or other geologic processes and events appears to be highly unlikely.

Yucca Mountain has a desert climate. This is important because water movement is the primary means by which radioactive waste could be transported from a repository. On average, Yucca Mountain currently receives about seven inches of rain and snow per year. Nearly all the precipitation, about 95 percent, either runs off or evaporates. Geological information indicates that the regional climate has changed over the past million years and the long-term average precipitation has been about 12 inches per year—comparable to that of present-day Santa Fe, New Mexico. Even if this were to be the case in the future, most of the water would run off or evaporate rather than soak into the ground and possibly reach the repository.

A repository would be built about 1,000 feet below the surface and 1,000 feet above the water table in what is called the unsaturated zone. The water table is about 2,000 feet beneath the crest of Yucca Mountain. Any precipitation that does not run off or evaporate at the surface would have to seep down nearly 1,000 feet before reaching the repository. Between the repository and the water table, it would have to move through another

1,000 feet of the unsaturated zone before reaching the water table. The groundwater in the region is trapped within a closed desert basin and does not flow into any rivers that reach the ocean.

The concept of disposing of waste in the unsaturated zone in the desert regions of the Southwest was first advanced by the U.S. Geological Survey in the 1970s. In 1976, the director of the Geological Survey suggested that the region in and around the Nevada nuclear weapons test site offered a variety of geologic formations and other attractive features, including remoteness and an arid climate.[15] In 1981, a Geological Survey scientist noted that the desert Southwest has water tables that are among the deepest in the world and that the region contains multiple natural barriers that could isolate wastes for "tens of thousands to perhaps hundreds of thousands of years."[16]

The Reference Design

In the current reference design, spent nuclear fuel and high-level radioactive waste would be transported to Yucca Mountain by truck or rail in specially designed, shielded shipping containers licensed by the Nuclear Regulatory Commission; removed from the shipping containers and placed in long-lived waste packages for disposal; carried into the underground repository by rail cars; placed on supports in the tunnels; and monitored until the repository is finally closed and sealed. Figure 4.8 shows the proposed plan for storage of high-level nuclear wastes at Yucca Mountain.

Surface facilities and operations. Surface facilities would be designed to receive the waste and prepare it for final disposal, and to support the excavation, construction, loading, and ventilation of the repository tunnels. The entire surface layout would cover about 100 acres and have three main areas:

At the north entrance to the underground repository would be the facilities and equipment to transfer waste from shipping containers to waste packages. Each waste package would be welded closed and thoroughly checked before being loaded onto a shielded transporter to be taken underground.

At the south entrance would be the facilities to support the excavation and construction of the tunnels. Near the top of the mountain would be the facilities that house the air intake and

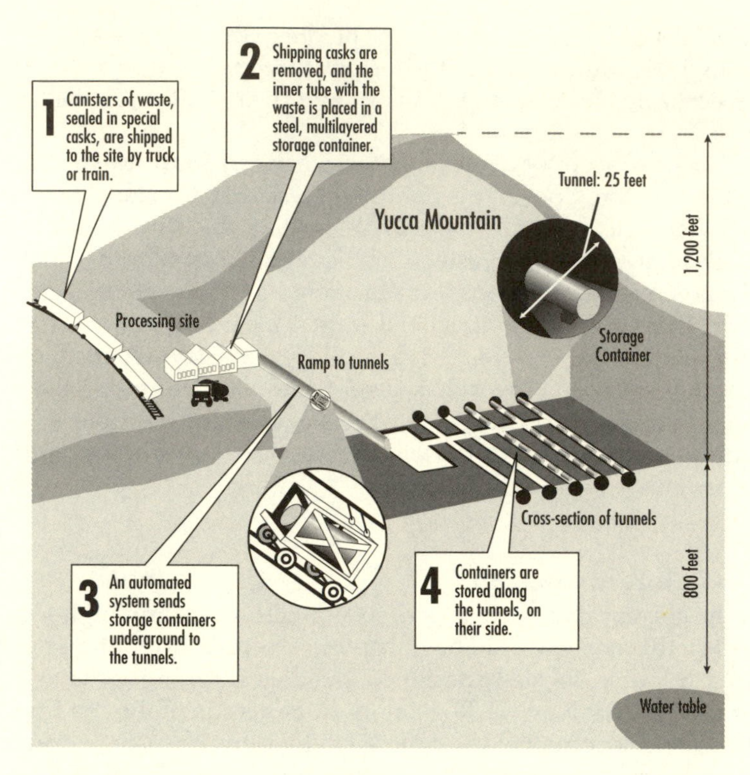

Figure 4.8 The Yucca Mountain disposal plan (Department of Energy and the Nuclear Energy Institute)

exhaust fans for ventilating the repository. Workers would be shielded from direct exposure to radiation and contamination because waste would be handled remotely.

Underground facilities and operations. The underground repository would consist of about 100 miles of tunnels. The main tunnels would be designed for moving workers, equipment, and waste packages. Ventilation tunnels would provide air for workers. The emplacement tunnels (or drifts) would accommodate the waste packages. Two gently sloping access ramps and two vertical ventilation shafts would connect the underground and surface areas.

Transportation underground would be by rail. A locomotive would haul the shielded transporter with its waste package underground from the waste-handling building to the entrance of an emplacement drift. Then a remotely operated crane (or gantry) would lift the waste package, carry it along the drift, and lower it onto its supports.

Current schedules anticipate that waste emplacement would begin in 2010 if a license is received from the Nuclear Regulatory Commission, after construction of surface facilities, the main tunnels, ventilation system, and initial emplacement drifts. Additional drifts would be constructed over a period of about 20 years while waste is being emplaced. The current design would accommodate 70,000 metric tons of waste, a limit imposed by the Nuclear Waste Policy Act of 1982. However, the site is large enough to accommodate additional waste, if that were authorized.

The engineered barrier system. The engineered barrier system is designed to work with the natural geologic barriers. The reference repository design features a long-lived waste package and includes the waste form, the concrete tunnel floor (or invert), and the steel and concrete support for the waste package.

The current waste package design would have two layers: a structurally strong outer layer of carbon steel nearly four inches thick, and a corrosion-resistant inner layer of a high-nickel alloy about three-fourths of an inch thick. These two layers would work together to preserve the integrity of the waste package.

The waste forms inside the waste package would provide additional barriers against transport of radionuclides away from the repository. Most spent nuclear fuel is encased in Zircaloy, a metal cladding that is highly resistant to corrosion. Defense high-level radioactive waste would be solidified as glass inside stainless steel canisters.

As the design process continues, DOE is evaluating several design options that might increase the ability of the engineered barrier system to contain waste. These include the following:

Drip shields that could keep water from dripping on the waste packages

Ceramic coating on the waste packages that could further prevent corrosion

Backfill that could protect the waste packages from falling rock or tunnel collapse, raise the waste packages'

> temperature and lower the relative humidity. Backfill would consist of crushed rock or other granular material that would be placed around the waste packages in the emplacement drifts just before the repository is closed.

The DOE also is evaluating alternative designs, some of which might reduce uncertainties regarding repository performance. . . .

Confirmation and retrieval. Activities to confirm that a repository would work as expected begin long before the first waste is emplaced. In the current site characterization phase, information about Yucca Mountain and the surrounding environment is being collected and compiled to provide a baseline against which to compare what would happen if a repository were built and waste were emplaced.

Using mathematical models based on the collected data and analyses of the engineered components, scientists forecast the probable behavior of the engineered system and the effects of a repository on the Yucca Mountain environment. If repository operations begin, remote sensors would monitor the waste packages, tunnels, and surrounding rock. The effects of a repository would be monitored, and the observed effects would be compared to the model predictions. These confirmation activities would help determine whether a repository is operating as expected.

If a problem is detected prior to closing the repository, remedial action or retrieval of the waste would be possible using remotely operated equipment. The Nuclear Regulatory Commission currently requires that a repository be designed to allow the retrieval of waste at any time up to 50 years after waste operations begin. Retrieval of waste, if needed, would follow, in reverse order, the same steps taken in emplacing the waste.

Repository closing. Even under the most ambitious schedules for disposal, future generations would make the final decision to close a repository. To give future generations the option of closing the repository or monitoring it for long periods of time, DOE is designing the repository so that it could (with Nuclear Regulatory Commission approval) be either closed as early as 10 years after emplacement of the last waste package, or kept open for hundreds of years from the start of waste emplacement.

Permanently closing the repository would require the sealing of all shafts, ramps, exploratory boreholes, and other underground openings. These actions would discourage any human intrusion into the repository and prevent water from entering through these openings.

At the surface, all radiological areas would be decontaminated, all structures removed, and all wastes and debris disposed of at approved sites. The surface area would be restored as closely as possible to its original condition. Permanent monuments would be erected around the site to warn any future generations of the presence and nature of the buried wastes.

The DOE also would continue to oversee the Yucca Mountain site to prevent any activity that could breach a repository's engineered or geologic barriers, or otherwise increase the exposure of the public to radiation beyond allowable limits. . . .

The Attributes of Safe Disposal

The results of fifteen years of testing and analysis, including four years of underground exploration, have validated many, but not all, of the expectations of scientists who first suggested that remote desert regions are well-suited for a geologic repository. One important and unexpected test result was finding underground, at the level of the proposed repository, traces of a radioactive isotope (chlorine-36) that is associated with above-ground nuclear weapons tests. As atmospheric nuclear testing began in the mid-1940s, this finding suggests that some water travels from the ground surface to the level of the repository in about 50 years or less. Another important finding was evidence that the average amount of water that filters down through the mountain is about a third of an inch per year, which, while only about five percent of the average annual precipitation, is more than DOE initially expected. Taken together, the findings, both expected and unexpected, underscore the importance of building engineered barriers that work with the natural barriers to keep water away from the waste.

The results indicate that a repository at Yucca Mountain would need to exhibit four key attributes to protect public health and the environment for thousands of years. The four key attributes are:

Limited water contact with waste packages

Long waste package lifetime

Low rate of release of radionuclides from breached waste packages

Reduction in the concentration of radionuclides as they are transported from breached waste packages

Based on performance assessment models, DOE has evaluated the degree to which the reference design exhibits these four key attributes, and has identified additional scientific studies and design improvements that could reduce uncertainties and enhance long-term repository performance.

Limited water contact with waste packages. In the reference design, waste packages would be placed about 1,000 feet below the mountain's surface and about 1,000 feet above the water table. Even if future climates are much wetter than today, the mountain is not expected to erode and leave the waste exposed, and the water table is not expected to rise high enough to reach the waste.

In the current semiarid climate, about seven inches of water a year from rain and snow fall on Yucca Mountain. Nearly all of that precipitation, about 95 percent, runs off or evaporates. Only about one-third of an inch of water per year moves down (or percolates) through the nearly 1,000 feet of rock to reach the level of the repository. Studies of past climates indicate that the precipitation may increase to a long-term average of about 12 inches per year. However, most of the water still would run off or evaporate rather than soak into the ground.

Once waste packages have been placed in the repository, the heat generated from radioactive decay would raise the temperature in the tunnels above the boiling point of water. The heat is expected to dry out the surrounding rock and drive any water away for hundreds to thousands of years. However, as the waste decays and the repository cools, enough water to cause drips would begin to seep into the drifts through fractures in the rock.

Using mathematical models, analysts estimate that, after the repository cools enough, about five percent of the packages could experience dripping water, under the current climate. If the climate changes to a wetter long-term average, about 30 percent of the packages could experience dripping water. These estimates are based on a number of assumptions that remain to be validated. Nonetheless, the results suggest that limited water would contact the waste packages.

Ongoing testing in the exploratory tunnels is providing more information on how much water could enter the repository and contact the waste packages under a variety of conditions. The DOE is also evaluating alternative waste package designs and other options that would mitigate the effects of water contact and improve performance of a repository.

Long waste package lifetime. The waste package in the reference design has two layers: a thick outer layer made of carbon steel that provides structural strength and delays any contact of water with the inner layer, and a thinner inner layer of a high-nickel alloy that resists corrosion after the outer layer is penetrated.

Based on preliminary results of corrosion experiments and the opinions of experts, computer simulations indicate that most of the waste packages would last more than 10,000 years, even if water is dripping on them. The longevity of man-made materials in the repository environment over such long periods of time is subject to significant uncertainty, however, and some waste packages could fail earlier. Scientists estimate that dripping water could cause the first penetrations—tiny pinholes—to appear in some waste packages after about 4,000 years. More substantial penetrations could begin to occur about 10,000 years later. Projections of waste package performance also assume that at least one waste package will fail in 1,000 years due to a manufacturing defect.

To reduce the uncertainty in waste package performance, further research on the conditions that waste packages will be exposed to and testing of waste package materials is underway. In addition, DOE is evaluating alternative waste package designs and materials that could compensate for the uncertainty and enhance longevity.

Low rate of release of radionuclides from breached waste packages. Once water enters a waste package, it would have to penetrate the metal cladding of the spent nuclear fuel to reach the waste. For about 99 percent of the commercial spent nuclear fuel, the cladding is highly corrosion-resistant metal that is designed to withstand the extreme temperature and radiation environment in the core of an operating nuclear reactor. Current models indicate that it would take thousands of years to corrode cladding sufficiently to allow water to reach the waste and begin to dissolve the radionuclides. However, estimates of cladding performance are uncertain, and more work in this area is planned.

During the thousands of years required for water to reach the waste, the radioactivity of most of the radionuclides would decay to virtually zero. For the remaining radionuclides to get out of the waste package, they must be dissolved in water, but few of the remaining radionuclides could be dissolved in water at a significant rate. Thus, only the long-lived, water-soluble radionuclides, such as isotopes of technetium, iodine, neptunium, and uranium, could get out of the waste package. Although most of the waste would not migrate from the package even if it were breached, the release of any radionuclides is reason for concern and motivation for seeking improvements in the repository design. Ongoing tests are providing more information on how radionuclides dissolve in water.

Reduction in the concentration of radionuclides as they are transported from the waste packages. Long-lived, water-soluble radionuclides that migrate from the waste packages will have to move down through about 1,000 feet of rock to the water table and then travel about 20 kilometers (about 12 miles) to reach a point where they could be taken up in a well and consumed or used to irrigate crops.

As the long-lived, water-soluble radionuclides begin to move down through the rock, some will stick (or adsorb) to the minerals in the rock and be delayed in reaching the water table. After reaching the water table, radionuclides will disperse to some extent in the larger volume of groundwater beneath Yucca Mountain, and the concentrations will be diluted. Eventually, groundwater with varying concentrations of different radionuclides will reach locations near Yucca Mountain where the water could be consumed.

Of the approximately 350 different radioactive isotopes present in spent nuclear fuel and high-level radioactive waste, six are present in sufficient quantities and are sufficiently long-lived, soluble, mobile, and hazardous to contribute significantly to calculated radiation exposures. Four of these isotopes—

technetium-99, iodine-129, neptunium-237, and uranium-234—can be transported by moving groundwater because they do not adsorb well to minerals. Two isotopes—plutonium-239 and plutonium-242—tend to adsorb but could be mobile because they can attach themselves to small particles (or colloids) and then be transported along with those particles.

Given the uncertainty about the rate at which groundwater moves and the possible existence of fast pathways or channels through the saturated zone, the DOE is continuing to investigate groundwater flow characteristics and is analyzing the possible effects on radionuclide transport and dilution.

Possible Dose

Analysts have calculated the possible radiation dose rate to people who may be living near the repository thousands of years in the future. Because where and how people will be living in the distant future cannot be predicted, analysts base their calculations on the current situation. They assume that the nearest population lives 20 kilometers (about 12 miles) from the repository boundary and has a lifestyle similar to the average person living today in Amargosa Valley, about 30 kilometers (about 19 miles) from Yucca Mountain.

During the first 10,000 years after the repository is closed, current models indicate that the mean peak annual dose rate to an average individual in this future population would be about 0.1 millirem. However, given the uncertainties associated with the assumptions and the performance assessment models, the peak dose could be higher or lower than the estimated average. There is a 5 percent (1 in 20) chance of exceeding 0.8 millirem and a greater than 25 percent chance of no exposure at all.

During the first 100,000 years, the mean peak annual dose rate to an average individual is estimated to be 30 millirem with a 5 percent chance of exceeding 200 millirem and a greater than 20 percent chance of zero dose.

During the first 1 million years, the mean peak annual dose rate to an average individual is estimated to reach 200 millirem, with a 5 percent chance of exceeding 1,000 millirem (or 1 rem) and a 5 percent chance of being lower than 0.07 millirem.

Other Safety Issues

The analysis of the safety of a repository at Yucca Mountain must also consider both the likelihood and the effect of possible disruptive processes and events, such as volcanism, earthquakes, human intrusion, and "nuclear criticality." The DOE has concluded that there is little likelihood that such processes or events at Yucca Mountain would significantly affect the long-term performance of a repository.

Volcanism. The area around Yucca Mountain was very active volcanically millions of years ago. The rock of Yucca Mountain—called tuff—is composed of volcanic ash from eruptions that occurred about 13 million years ago. However, large-scale volcanism in the area ceased about 7.5 million years ago, and the last, small eruption occurred about 75,000 years ago. Experts have concluded that the chance of future volcanic activity disrupting the site is negligible. As a result, volcanism would be unlikely to affect the long-term performance of the repository.

Earthquakes. Yucca Mountain is located in the southern Great Basin, a large region that has some earthquakes. Yucca Mountain itself is a tilted block of rock that is bounded by geologic faults. A magnitude 5.6 earthquake occurred about 12 miles away in 1992. A repository and surface facilities would be designed to withstand earthquakes, as are modern tunnels, buildings, and power plants in seismically active areas.

Accidental human intrusion. It is possible that future human activities might intrude on the repository. One possible activity would be exploration for valuable natural resources. However, Yucca Mountain exhibits few characteristics that would make it an attractive location for future generations to drill or otherwise explore for gold, hydrocarbons, or other materials.

The National Academy of Sciences (NAS) concluded that there is no scientific basis for predicting such human activities over the very long periods of time for which the repository must function. The NAS, therefore, recommended that future human intrusion not be considered in the quantitative performance assessments. However, to evaluate how the repository would perform if humans were to intrude, the NAS recommended,[20] and DOE has conducted, a separate analysis of a theoretical case in which a waste package is penetrated by someone drilling into the repository in the future. Performance assessments indicate that peak dose rates would increase if a waste package were penetrated by exploratory drilling and if waste were then carried down the drillhole to the water table. However, as noted, natural resource assessments indicate that the Yucca Mountain site does not exhibit characteristics that would make it an attractive location for exploratory drilling.

Nuclear criticality. A nuclear criticality occurs when sufficient quantities of fissionable materials come together in a precise manner and the required conditions exist to start and sustain a nuclear chain reaction. The waste packages would be designed to prevent a criticality from occurring inside a waste package. In addition, it is very unlikely that a sufficient quantity of fissionable materials could accumulate outside of the waste packages in the precise configuration and with the required conditions to create a criticality. If, somehow, an external criticality were to occur, analyses indicate that it would have only minor effects on repository performance. An explosive external criticality is not credible.

2. For a description and discussion of radioactive waste and its management, see The League of Women Voters 1993. *The Nuclear Waste Primer: A Handbook for Citizens.* New York: League of Women Voters Education Fund. 210697.

3. U.S. District Court, Utah 1995. Joint Motion for Entry of Consent Order Based on Settlement Agreement and Consent Order in the Case of Public Service Co. of Colorado v. Batt, October 17, 1995. Civil Case No. 91-0054-S-EJL (Legal Pleadings). U.S. District Court for the District of Idaho. 240346.

4. U.S. Department of Energy 1997. *Linking Legacies: Connecting the Cold War Nuclear Production Processes to Their Environmental Consequences.* DOE/EM-0319, pp. 34–38. Washington, D.C.: DOE. 241255.

10. Nuclear Waste Policy Amendments Act of 1987. Public Law 100-203. 223717.

11. 10 CFR 60. Energy: Disposal of High-Level Radioactive Wastes in Geologic Repositories. 239474.

12. Energy Policy Act of 1992. Public Law 102-486. 233191.

13. National Research Council 1995. *Technical Bases for Yucca Mountain Standards.* Washington, D.C.: National Academy Press. 104273.

14. Interagency Review Group on Nuclear Waste Management 1979. *Report to the President by the Interagency Review Group on Nuclear*

Waste Management. TID-29442, p. 37. Washington, D.C.: DOE. MOL.19980625.0169.

15. Letter from Dr. Vincent McKelvey to Richard W. Roberts, Assistant Administrator for Nuclear Energy, U.S. Energy Research and Development Administration, Washington, D.C. July 9, 1976. MOL.19990119.0314.

16. Winograd, I. J. 1981. "Radioactive Waste Disposal in Thick Unsaturated Zones." *Science, 212,* pp. 1457–1464. Washington, D.C.: American Association for the Advancement of Science. 217258.

20. National Research Council 1995. *Technical Bases for Yucca Mountain Standards,* p. 12. Washington, D.C.: National Academy Press. 104273.

Source: Department of Energy, "Yucca Mountain Viability," December 1998, excerpts, with illustrations omitted.

Opposition to Nuclear Power

A number of groups have expressed opposition to nuclear power and have released statements challenging various aspects of the technology on safety, environmental, and sociopolitical grounds. Following are some representative statements from antinuclear groups.

Nuclear Power's Failed Promise

In the 1960s the nuclear power industry promised Americans "energy too cheap to meter." Instead the American consumer has been saddled with one of the most complicated and costly means of boiling water yet devised. Contrary to the claims made by the nuclear propagandists, nuclear power has proven to be unsafe, uneconomical, and unnecessary.

Nuclear Power & Atomic Accidents

The meltdown at Three Mile Island and the explosion at Chernobyl irreparably altered the image of nuclear power in the United States and around the world. The dramatic decrease in nuclear construction can be directly tied to the meltdown at Three Mile Island. The horrific images of the Chernobyl disaster and the ever-growing death toll are a constant reminder of the dangers of nuclear power. In 1985, the U.S. Nuclear Regulatory

Commission testified to Congress that the risk of a meltdown at a U.S. reactor was 45% in the next 20 years.

Nuclear Power's Economic Meltdown

However, the risks of nuclear power are only part of the problem. It has been the nuclear industry's inability to manage the construction and operation of its nuclear reactors that has solidified public opposition to nuclear power in the United States.

Chronic escalation of construction costs coupled with high operation and maintenance costs have sealed nuclear power's economic fate. When construction costs skyrocketed and O&M costs spiraled out of control, nuclear power became an economic disaster. In 1986, the U.S. Department of Energy compared nuclear construction cost estimates to the actual final costs for 75 reactors. The original cost estimate was $45 billion. The actual cost was $145 billion! *Forbes* magazine recognized that this "failure of the U.S. nuclear power program ranks as the largest managerial disaster in business history, a disaster of monumental scale." According to *Forbes,* "only the blind, or the biased, can now think the money has been well spent."

Today, forty-two of the 105 nuclear plants in the U.S. have operations and maintenance costs that are greater than the price of replacement power. Rather than energy "too cheap to meter," nuclear reactors have resulted in some of the highest electricity rates in the nation. Reactors have bankrupted a number of utilities and driven others to the brink. The shift from a regulated to a deregulated market for electricity will place even greater economic pressure on utilities that operate nuclear reactors. The Arthur Anderson consulting firm concluded that "the threat exists that nuclear utilities in their desire to cut costs and increase competitiveness will be forced to impair their operational safety and increase risk."

The Decline and Fall of Nuclear Power

Nuclear power is in decline in the United States. U.S. utilities have canceled almost as many nuclear reactors as they have constructed. No nuclear reactors have been ordered and subsequently completed in the U.S. since 1973. The last nuclear reactor to be constructed in the United States was completed in 1996; the Tennessee Valley Authority's Watts Bar reactor took almost 23 years to build and cost nearly $8 billion. By 2035, every nuclear reactor currently licensed to operate in the U.S. will be shut down.

Impending deregulation in the electricity industry has nuclear utility executives in a quandary. Their nuclear reactors are too expensive to operate and too costly to shut down. Regardless of the shortfall in decommissioning funds a number of utilities have decided to send their nuclear reactors into early retirement.

Not only are nuclear reactors being shut down, but the "next generation" of reactors may never be built in the U.S. In an annual survey of U.S. utility executives, the Washington International Energy Group found that 89% of respondents stated that they would not order a new nuclear reactor.

Nuclear power is an inherently dangerous activity and has proven to be a very expensive way to boil water. Even if it were possible to build a safe and economic reactor, the nuclear industry would still be faced with the intractable problem of long-lived radioactive wastes. Nothing should be done to prevent the demise of nuclear power. It is a failed experiment and deserves no further support from U.S. taxpayers.

Nuclear Power & Global Warming

Nuclear power has no role to play in combating greenhouse emissions. Although nuclear reactors produce no greenhouse gases, replacing coal-fired generation with nuclear reactors is not politically or economically feasible. To have any substantial impact upon greenhouse gas emissions, nuclear-generated electric power must be affordable and politically palatable. It is neither.

Almost ten years ago, the Rocky Mountain Institute (RMI) debunked the nuclear industry's claims that nuclear reactors could combat global warming. They developed a scenario that would reduce greenhouse emissions by 20 to 30 percent by 2050 by replacing coal fired power plants with nuclear reactors. The analysts at RMI found that such a scenario would require the completion of a nuclear reactor every one to three days for the next 40 years. The total cost would run as much as $9 trillion.

Clearly, nuclear power is too expensive and has too long a lead time to effectively counter CO_2 emissions.

Alternatively, the analysts from RMI found that every dollar spent on energy efficiency is approximately seven times more effective in reducing CO_2 than a dollar spent on nuclear power.

Nuclear Power & Nuclear Weapon Proliferation

Any further expansion of commercial nuclear power carries with it the risk of nuclear weapons proliferation. From the dawn of the nuclear age it has been recognized that nuclear power and nuclear weapons were inextricably linked. The 1946 Acheson-Lilienthal report on the control of atomic energy recognized that, "the development of atomic energy for peaceful puposes and the development of atomic energy for bombs are in much of their course interchangeable and interdependent." It is no coincidence that the U.S., the Soviet Union, France and Great Britain all developed commercial programs in conjunction with their bomb building efforts. Civilian programs sprang from the perceived need to produce plutonium.

Civilian nuclear programs have led to the proliferation of nuclear weapons in India, Pakistan, Israel, and South Africa. Many commercial nuclear aspirations in the developing world are viewed as slightly veiled attempts to secure nuclear weapons.

The U.S. government should severely limit the export of nuclear technology. While U.S. corporations like General Electric and Westinghouse attempt to sell their reactors abroad, they undermine the nonproliferation policies and ultimately the national security of the United States. In order to ensure the maximum decoupling of nuclear power and nuclear weapons, the U.S. should first lead by example. The ill advised plans of the Department of energy to produce tritium in commercial reactors should be scrapped. It is hypocritical and weakens our nonproliferation policy.

Source: Fact sheet, Public Citizen/Critical Mass Energy Project

The Nuclear Waste Controversy

A number of antinuclear groups (and some scientists in national laboratories) have raised concerns about the handling of nuclear waste in general and the design of the Yucca Mountain facility in particular. Here are two representative statements.

What's Wrong with Burying Nuclear Waste at Yucca Mountain?

In 1987 Congress selected Yucca Mountain in Nevada as the sole candidate to be studied for a permanent repository for high-level nuclear waste from the nation's commercial reactors. Unfortunately, the decision was based more on political expediency than scientific consensus. The Department of Energy expects to assess Yucca Mountain's viability for a repository in 1998, followed by a site suitability determination of Yucca Mountain in 2001. The viability assessment is a preliminary engineering and cost assessment. The site suitability determination is the decision that a license application will be filed with the Nuclear Regulatory Commission to construct a repository within Yucca Mountain. Since the mountain's selection as the sole candidate, site suitability studies have raised serious technical questions and the program has provoked extensive criticism.

An Unsuitable Site

The toxic materials in irradiated reactor fuel will remain lethal for hundreds of thousands of years. In the early 1980s, burial of the high-level waste was seen by many as the best option for disposal. Since then, however, complex and significant doubts have been raised about a geologic repository's ability to ensure the irradiated fuel's isolation. Yucca Mountain in particular has numerous specific features that make the site unsuitable for the task.

Earthquakes. At least 33 known earthquake faults lie in Yucca Mountain's vicinity. Studies by the Geological Survey discovered that the Ghost Dance Fault, which crosses the site, may be the primary fault of a complex fault zone.[1] The area has had 621 seismic events of magnitude greater than 2.5 within a 50 mile radius of Yucca Mountain within the last 20 years.[2] In 1992 an earthquake that registered 5.6 on the Richter scale occurred 12 miles away and damaged a DOE field office building. The Nuclear Waste Technical Review Board (NWTRB), an advisory body established by the Nuclear Waste Policy Act to monitor the waste program, also warns that extensive fault systems may not leave sufficient emplacement space for nuclear waste.[3]

Another danger from the region's seismic activity involves the water table, which is 300 meters below the proposed repository. Former senior Department of Energy geologist Jerry Szymanski has found that an earthquake could dramatically elevate the table, flooding the repository with water.[4] In a study to be released in August 1997, researchers from the University of Colorado at Boulder concur with this theory. A magnitude 5 or 6 earthquake could raise the water table 450 to 700 feet, flooding the repository. John B. Davies, a physics research associate with the study, says, "If water hits the storage area it could cause a rapid corrosive breakdown of the containers and allow the plutonium to leak into the water table and the atmosphere."[5]

Water. The movement of water through the site also represents a serious threat to a repository. One of Yucca Mountain's supposed advantages is slow travel time of the water through the ground. Studies suggest, however, that water may move through the mountain at rates faster than once thought.[6] The extensive fault system in the area also creates a risk that pathways may be created through Yucca Mountain's highly fractured rock for water to reach the repository directly. A March 1996 study by Los Alamos National Laboratory found rain water at the proposed repository site that is only 40 years old.[7] This evidence conflicts sharply with previous estimates that it would take thousands of years for water to seep down to that depth.

Nevada scientists are also concerned that a repository may lead to groundwater contamination, fearing that groundwater travel time from the repository to the environment is less than 1,000 years, instead of the many thousands of years that DOE claims.[8] The GAO has found that limitations in the data used by DOE for assessing groundwater flow rates are inadequate. A DOE report, cited by the GAO, on the quality of the Geological Survey's hydrologic investigations found that "major uncertainties, such as the unexplained drop in the ground water level, at this stage of the scientific investigation limit understanding of how radioactive materials would move in groundwater."[9]

Volcanic activity. A volcano 20 kilometers away from the site appears to have erupted within the last 20,000 years, rather than 270,000 as once thought.[10] The interval becomes comparatively small when one remembers that the materials to be buried at Yucca will remain highly toxic for a quarter of a million years.

Defense wastes. In addition to holding irradiated fuel from commercial reactors, Yucca Mountain is the intended destination for certain defense wastes in the form of vitrified borosilicate glass. Vitrified glass, however, may disintegrate rapidly in conditions like those at Yucca Mountain, which may result in massive groundwater contamination.[11]

Criticality. A team of scientists at the Los Alamos National Laboratory in New Mexico fear that burying waste at Yucca Mountain could lead to a spontaneous atomic explosion, releasing radiation into the atmosphere and groundwater. According to their analysis, plutonium could escape from disposal canisters into the surrounding rock, which possesses physical properties that might aid a spontaneous chain reaction and explosion. Another team of scientists at the DOE facility at Savannah River have endorsed this thesis.[12]

Human intrusion. Yucca Mountain is located in an area rich with mineral resources, which may lead to human intrusion upon the site after waste has been deposited.[13]

Political Considerations

Federalism/state authority. Although Nevada has no nuclear reactors, Congress chose the state to be the only candidate for a permanent repository. Nevada has a long history of vigorous opposition to the repository. In 1989 the state passed a law to prohibit the storage or disposal of high-level nuclear waste within Nevada, and public opinion polls consistently show strong citizen disapproval of the dump. While current law prohibits the placing of an interim facility for high-level nuclear waste in Nevada, legislation now before Congress would circumvent the restriction and allow the preemption of Nevada state laws, raising issues of federal preemption and state sovereignty.

Legacy of doubt. One legacy of the DOE's handling of the characterization study (see below) is deep mistrust of the department by residents of Nevada. While considerable anger remains from the manner in which Yucca Mountain was selected as the only candidate for a repository, much citizen opposition stems from the realities of nuclear waste disposition. As one DOE-commissioned report observes, "[p]ublic mistrust of DOE's nuclear waste storage program is sometimes rooted in irrational fear of unknown risk to health and safety, but more often reflects a rational understanding of the current state of scientific knowledge and of DOE's past history of covering up mistakes and censoring bad news."[14]

Native American sovereignty. The federal government may lack authority to even use Yucca Mountain. The site lies in an area that is part of the traditional lands of the Western Shoshone tribe. The Shoshone nation claims the area and vicinity under the Treaty of Ruby Valley, signed in 1863, and opposes the DOE's plans for Yucca Mountain.[15]

DOE management at Yucca: A legacy of failure. The nuclear industry is backing legislation to explicitly waive environmental regulations for a repository. Even without such legislation, however, past mismanagement and cost overruns call into question the integrity and effectiveness of the site suitability studies. Numerous observers have taken the DOE to task for its past management of the study.

Project integrity. Nevada's elected representatives have long accused the DOE of failing to conduct an honest evaluation of Yucca Mountain, but even individuals employed by the department have come to the same conclusion. Retired Air Force Brigadier General Joel T. Hall, employed by a DOE contractor to determine ways to improve the credibility of the Yucca Mountain program, wrote a letter to then–DOE Secretary James Watkins and accused the Department of using its studies not for site suitability research, but as a prelude to applying for a construction license. "It is a sad day for our country when the public becomes unjustly cynical about the integrity of public officials," he wrote. "But, it is so much sadder when the cynicism is justified. The Yucca Mountain Project falls in the latter category."[16]

Money down the hole. The poor management of the site characterization presents a financial drain as program costs continue to escalate. Characterization of Yucca Mountain, once estimated in the hundreds of millions, may cost in excess of $6 billion.[17] Most of these payments come from the Nuclear Waste Fund, which is financed by a tenth of a cent per kilowatt hour fee assessed to nuclear utilities. The waste fund is likely to prove inadequate because the fee has not changed since its establishment in 1983. Up to 1995, inflation eroded the fee's buying power by 45 percent.[18] Although the fund has collected in excess of $10.8 billion, over $4.8 billion has already been spent,[19] including $182 million in FY'97.

Continuing problems and new misdirections. While the DOE claims to be instituting management reform of the project, thorough site suitability analysis continues to be sacrificed on the altar of expediency. DOE's "program approach" in many ways worsens the problem of data collection. In March 1995 the NWTRB expressed concern that "increased technical and scientific uncertainties will be created because less data and analysis than previously planned will be available prior to determining site suitability."[20] NWTRB Chairman Dr. John E. Cantlon testified that the existing schedule may be inadequate for site exploration or the accommodation of unforeseen circumstances. "The Board is very concerned that important program decisions are being driven by unrealistic deadlines."[21]

The deferral of important studies has already drawn the attention of affected parties outside of Nevada. In a letter to then DOE Secretary Hazel O'Leary, dated January 31, 1995, Senators Barbara Boxer and Dianne Feinstein raised questions about the effect of the new approach on understanding how a repository at Yucca Mountain might affect the Death Valley National Park. "We are concerned that, while earlier plans indicated an effort would be made to define and analyze the regional groundwater flow system . . . the new program approach defers development of an understanding of the regional ground-water flow system until after the technical determination of site suitability."

Another continuing problem is the failure to fully fund scientific research of the site's suitability. The General Accounting Office has commented that the DOE devotes most project funds to infrastructure and a relatively small amount to essential characterization studies. At the same time, the agency reduced the time allotted for various studies, a move which "could increase the risk that the site investigation will be inadequate and comes at a time when unanticipated technical issues have emerged that could lengthen the investigation."[22]

A GAO report reviewing ethical questions at Yucca Mountain has found that both the Manager (1987–1993) and the Deputy Manager (1990–1994) of DOE's Yucca Mountain Project "had long-term personal relationships with personnel of major project contractors." In addition, "DOE has learned that 14 additional, or almost 18 percent of, DOE employees at the project were engaged in relationships that might have created problems concerning the lack of impartiality and independence."[23]

Studies deferred. In March 1995 the NWTRB expressed concern that the DOE was delaying important studies of areas of the site until after the determination of viability. The Board also noted the need for more data on how Yucca Mountain's rock would respond to heat generated by nuclear waste. "Unfortunately, no hydrogeologic data from thermal testing in the exploratory facility will be available to support a 1998 viability assessment. The Board is very uneasy about this."[24] The Board further noted that the DOE's preparation of an Environmental Impact Statement for the repository may not place sufficient emphasis on the release of radiation into the environment.[25]

The DOE is also altering its plans at Yucca in a manner that may cast further doubt upon the site suitability decision. The NWTRB expressed concern that while the DOE had placed an emphasis on natural geologic barriers to prevent nuclear waste from entering the environment, the department appears to have shifted its focus to engineered barriers as the waste's primary containment, using geologic barriers as a backup. The effectiveness of this backup, however, will take longer to demonstrate. The Board questioned whether the new strategy will provide adequate protection and if demonstration of geologic isolation could wait until after site suitability had been determined.[26]

Calls for an independent review. Independent bodies like the General Accounting Office and the Nuclear Waste Technical Review Board have consistently called for an independent review of the Yucca Mountain project and, in the case of the GAO, of the entire program. "Without a comprehensive independent review of the disposal program and its policies," warns the GAO, "millions—if not billions—of dollars could be wasted."[27]

1. The General Accounting Office, *Nuclear Waste: Yucca Mountain Project Behind Schedule and Facing Major Scientific Uncertainties* (GAO/RCED-93-124), p. 43.

2. Council of the National Seismic System Composite Catalog, 1976 to present; and Southern Great Basin Seismic Network.

3. Enclosure of letter from John E. Cantlon, Chairman NWTRB, to Daniel A. Dreyfus, Director OCRWM, December 6, 1994.

4. Nicholas Lenssen, "Nuclear Waste: The Problem That Won't Go Away." Worldwatch Paper 106, December 1991, p. 17.

5. Davies and Archambeau, Press Release, June 24, 1997, University of Colorado at Boulder.

6. Keith Rogers, "Policy Holds No Alternative to Yucca Mountain," *Las Vegas Review-Journal.* October 13, 1994, p. 4A; Lester R. Brown, et al., *Vital Signs 1995: The Trends That Are Shaping Our Future* (Worldwatch Institute/W.W. Norton: New York, 1995), p. 88.

7. Matthew L. Wald, "Doubt Cast on Prime Site as Nuclear Waste Dump," *New York Times,* June 20, 1997, p. A12.

8. State of Nevada Nuclear Waste Project Office, "Scientific and Technical Concerns," p. 14.

9. General Accounting Office, *Nuclear Waste: Impediments to Completing the Yucca Mountain Repository Project.* RCED-97-30. January 1997, p. 13.

10. Lenssen, 27.

11. Arjun Makhijani, "Glass in the Rocks: Some Issues Concerning the Disposal of Radioactive Borosilicate Glass in a Yucca Mountain Repository," prepared for the State of Nevada, January 29, 1991.

12. William J. Broad, "Scientists Fear Atomic Explosion of Buried Waste," *The New York Times,* March 5, 1995, p. 1; William J. Broad, "Theory on Threat of Blast at Nuclear Waste Site Gains Support," *The New York Times,* March 23, 1995, p. A18.

13. Scott Saleska, *Nuclear Legacy: An Overview of the Places, Problems, and Politics of Radioactive Waste in the U.S.* (Public Citizen, September 1989), p. VII-7.

14. James Thurber, *Report on Selected Published Works and Written Comments Regarding the Office of Civilian Radioactive Waste Management Program.* Center for Congressional and Presidential Studies, School of Public Affairs, The American University (Washington, March 1, 1994), p. 17.

15. Memorandum from Ian D. Zabarte, Manager Western Shoshone National Council–Nuclear Waste Program, May 16, 1994.

16. Letter from Joel T. Hall to James D. Watkins, April 22, 1992, p. 7.

17. U.S. Department of Energy, *Civilian Radioactive Waste Management Program Plan: Volume I, Program Overview* (Washington, December 19, 1994), p. 6.

18. Congressional Budget Office, *Reducing the Deficit: Spending and Revenue Options.* February 1995, p. 233.

19. Data (May 31, 1997) provided by Charlie Smith of the DOE Budget Office, July 15, 1997.

20. Written testimony of Dr. John E. Cantlon, Chairman Nuclear Waste Technical Review Board, before the Subcommittee on Energy and Water Development of the House Committee on Appropriations, March 16, 1995, p. 7.

21. Ibid., p. 11.

22. The General Accounting Office, *Nuclear Waste: Yucca Mountain Project Behind Schedule and Facing Major Scientific Uncertainties* (GAO/RCED-93-124), p. 3.

23. The General Accounting Office, *Department of Energy: Unethical Conduct at DOE's Yucca Mountain Project* (GAO/OSI-96-2), p. 1.

24. Letter from John E. Cantlon, Chairman NWTRB, to Daniel A. Dreyfus, Director OCRWM, March 3, 1995.

25. Ibid.

26. Ibid.

27. General Accounting Office, *Nuclear Waste: Comprehensive Review of the Disposal Program Is Needed* (GAO/RCED-94-299), p. 2.

Source: Public Citizen/Critical Mass Energy Project, July 1997

Why We Call It "Mobile Chernobyl"

The proposed Yucca Mountain, Nevada nuclear waste repository and proposed centralized storage of waste at that site would trigger the largest nuclear shipping campaign in history. 43 states would be run over by thousands of nuclear waste shipments (truck and train). 50 million people live within 1/2 mile of the projected routes.

The Yucca site has been targeted for a permanent nuclear waste repository since 1987. In the mid 1990's the nuclear utilities, the primary producers of the waste, decided to try and change existing law so that waste could be shipped to the site immediately for storage. This legislation has been dubbed the "Mobile Chernobyl" bill.

The legislation was initially introduced by the nuclear industry's Congressional champions in 1994. The environmental and consumer advocates groups and the Clinton administration decided that this was a bad idea. In fighting the proposal, it has been useful to "call a spade a spade."

Why is the largest nuclear waste shipping campaign in history legitimately named for the worst nuclear reactor accident to date?

This high-level nuclear waste, also called irradiated fuel or the misleading industry term "spent fuel," is mostly the fuel from a commercial nuclear power reactor; the same material that was scattered by the Chernobyl accident.

The Chernobyl reactor exploded on April 26, 1986 and then burned for days before it was extinguished. During that time particles of highly radioactive irradiated fuel were lofted into a plume which then made "fall-out" locally and over much of Europe. The radioactivity eventually circled the Northern Hemisphere six times, and measured here [in the United States].

One of the **worst-case scenarios for a transport accident** with irradiated fuel is where the shipping container is compromised—perhaps only partially cracked open . . . and is engulfed in flames. In this case it would not be a nuclear fire, but a high-temperature diesel fire would loft particles of irradiated fuel just like the Chernobyl plume, though smaller. Nonetheless, those affected would be affected in much the same way that Chernobyl has impacted millions of people, [and] soil, water, food, animals. . . .

Particles of irradiated fuel can cause a lethal exposure if concentrated—such as workers, motorists and emergency responders might face at the accident site. Many of the workers and clean-up workers from Chernobyl have died. **In lower concentrations, this contamination will cause additional cancers (both fatal and non-fatal), birth defects, genetic defects, diseases and disorders associated with lowered immunity and sterility.**

Thus a really bad transport accident with irradiated fuel has the capacity to cause permanent and ongoing impacts to the environment, to people, to resources, to property.

The Department of Energy (DOE) may be contracting the transport of this deadly cargo to **private contractors** on the basis of a fixed price contract. This means that the contractor makes their profits by keeping the costs low. At the same time, DOE will offer complete indemnification for the contractor, removing any incentive to be sure that extra effort is put into safety, equipment or procedures that might take more time and cost more money but would lower risk and/or hazard of transporting nuclear waste.

Some people say that it is not accurate to use the name Chernobyl since the waste in the container is not the same as an actively fissioning (splitting atoms) reactor. In fact, much care has to be taken to prevent nuclear waste from "going critical" and resuming the nuclear fission reaction. While it is possible to do this, the task is monumental. This is because each and every fuel rod is different. A reactor core is like an oven of sorts—more fission in the middle, less around the edges. Thus, each rod has a unique profile because of its position while it was in the reactor core. This results in variation in how much uranium and plutonium is present that could "go critical."

Therefore the problem of preventing criticality in a nuclear shipping cask, or a repository cask, for that matter, is one of bookkeeping. Each has to be 100% within the margin to prevent critical mass. As everyone knows, bookkeeping is subject to human error. What will be the margin of error on loading more than 10,000 containers of this deadly waste?

When the Department of Energy looks at accident rates they include many assumptions. The fact is, there will be accidents. Probably between 200 and 350 over the course of the program if Yucca Mountain is selected. This is because of the massive number of shipping miles. The average distance is about 2000 miles from where the waste is now to the Western Shoshone Land where Yucca Mountain sits. A DOE engineer stated in 1994 that he expects there will be 4–6 accidents that involve the release of radioactivity off the site.

When asked how this squares with the statement in DOE's Environmental Assessment of a Yucca Mountain repository that there would be "no significant radiological impact" from transport to the repository, the same engineer explained that since it was a national program, the radiation exposures were AVERAGED ACROSS THE ENTIRE US POPULATION. Thus the exposure of communities to irradiated fuel particles was not considered "significant."

Is it acceptable to you to be sacrificed for the sake of the Department of Energy's need to fulfill contracts they signed with the nuclear utilities and then averaged with millions who were not impacted? Is your community "significant" if it is permanently contaminated?

A second proposal that would also trigger "Mobile Chernobyl" is a "private" utility sponsored centralized interim storage site in Utah located on Skull Valley Goshute land. Southern Company is a key partner in developing this site.

THE ONLY CURE FOR NUCLEAR WASTE ACCIDENTS IS PREVENTION.

WORK TO OPPOSE MOVING NUCLEAR WASTE UNTIL WE ARE SURE IT WILL IMPROVE THE PROBLEM, NOT MAKE IT WORSE!!!!

Source: Nuclear Information and Resource Service (NIRS)

Legal Aspects of Nuclear Power

The regulation of nuclear power is a very complex field: Many aspects of the technology must be addressed by regulators, including operating standards, safety systems, environmental impact, worker safety, and public safety. This section presents a basic overview of the regulatory and legal arena surrounding nuclear power.

Regulations

The following laws have created the basic regulatory structure for nuclear power in the United States.

Atomic Energy Act of 1946 (60 Stat. 755, Former 42 USCA Sections 1801–1819)

This law reflected the urgent need to control the use of atomic weapons and energy following World War II. The law focused on the need to protect national security by keeping tight control over the production, acquisition, and distribution of fissionable materials. A new agency, the Atomic Energy Commission (AEC), was created for this purpose. Included in the agency's charter was the development of a program for the peaceful use of nuclear energy, especially for power production, and the development of a system for licensing and regulating private nuclear facilities.

Atomic Energy Act of 1954 (69 Stat. 919, USCA Sections 2011 et seq.)

As nuclear technology rapidly progressed to the point where private ownership and operation of nuclear reactors became feasible and economically attractive, there was a need to revamp the somewhat rudimentary 1946 legislation. The 1954 act improved both industry regulation and provisions for cooperation with U.S. allies. The legislators also sought to increase public awareness of the potential of nuclear power and the dissemination of information about nuclear energy.

The act includes provisions for the Atomic Energy Commission's role in overseeing utility contracts and procedures for the production, distribution, acquisition, and ownership of nuclear fuel, including mining and processing facilities. It also established the Joint Committee on Atomic Energy, which was later abolished.

Price-Anderson Act of 1957 (P.L. 85-256, 71 Stat. 576, 42 USCA Section 2210)

Because of uncertainty about the safety of nuclear energy and the possibility of a major accident causing considerable loss of life and potentially billions of dollars in property damage, insurance companies were reluctant to provide nuclear power plants with coverage sufficient to deal with such a disaster. This law therefore set an upper limit of $500 million for compensation for personal damages in nuclear accidents and established a fund of more than $500 million in government money, with utilities making a contribution to the fund. In 1988 the limit per accident was raised to $7 billion.

This provision of what amounts to "no-fault insurance" for nuclear plant operators has been strongly objected to by antinuclear activists, who believe it represents an unwarranted government subsidy to an industry that should not be allowed to operate if it is not safe enough to acquire insurance in the normal way.

Private Ownership of Special Nuclear Materials Act of 1964 (P.L. 88-489, 78 Stat. 602, 42 USCA Sections 2012 et seq.)

This law was based on both the importance of the growing nuclear power industry and the need to regulate the processing and use of nuclear materials such as uranium or plutonium fuel, and the byproducts of reactor operation. The act has provisions for compensating for damages caused by nuclear facilities in one state in another state, which the federal government regulated as an aspect of interstate commerce.

Energy Reorganization Act of 1974 (P.L. 93-438, 88 Stat. 1233, 42 USCA Sections 5801 et seq.)

The early 1970s saw growing public concern with the security of America's energy supply, especially in the light of the "oil shock" that led to long lines at gas pumps. The Energy Reorganization Act made important changes in the federal role in energy in general and nuclear energy in particular. The law established the Energy Research and Development Administration (ERDA), charged with funding research into energy efficiency and alternative forms of power. The former Atomic Energy Commission was split into two new agencies, the independent Nuclear Regulatory Commission, appointed by the president, and the Energy Research and Development Council.

Uranium Mill Tailings Radiation Control Act of 1978 (P.L. 95-604, 92 Stat. 3021, 42 USCA Sections 2014, 2021)

The mining of uranium creates leftover material called mill tailings. This material, which has low-level radiation (about 85 percent of that found in natural uranium ore), has been used in construction. However, as evidence of radiation health effects from even low doses of radiation mounted, Congress sought a safer way to dispose of mill tailings. The law created a program to "stabilize" disposal sites to prevent dispersal of tailings into the surrounding environment. Ninety percent of the costs would be paid by the federal government, with the remaining 10 percent being the responsibility of the states.

Low Level Waste Policy Act of 1980 (P.L. 96-573, 94 Stat. 3347, 42 USCA Sections 2014, 2021)

Concern with nuclear wastes continued with the passage of a law regulating the disposal of low-level nuclear wastes. The main problem with these wastes is that although they are not highly radioactive, they are produced in a great volume by nuclear plants as well as by industrial and medical facilities that use radioactive materials. There have been a number of cases of illegal or unsafe dumping of these wastes, and states and communities have proven reluctant to allow construction of a waste dump in their area. This law made it the policy of the United States that states would have primary responsibility for disposing of low-level waste, subject to federal regulations. Amendments added in 1985 provided incentives and further requirements that states (or regional compacts of states) provide adequate low-level waste facilities.

Nuclear Waste Policy Act of 1982 (P.L. 97-425, 96 Stat. 2201, 42 USCA Section 10101)

During the early 1980s efforts to create permanent nuclear waste sites were stalemated because the government, the nuclear industry, and environmental groups could not agree on a waste disposal policy. The Nuclear Waste Policy Act of 1982 established a timetable for creating a permanent underground repository for nuclear wastes. It required the Department of Energy to select five possible sites for the facility and to recommend three of them to the president by January 1, 1985. The president was to submit his choice for the first site to Congress by 1987, and the site was to receive nuclear waste no later than January 31, 1998.

This timetable has not been met. However, in 1996 the Department of Energy began a viability assessment of one site, Yucca Mountain in Nevada. A decision is expected in 2001 whether to recommend to the president that the site be developed.

International Cooperation and Treaties

The following U.S. laws and international treaties have specified cooperation in the development of nuclear power or in the prevention of nuclear proliferation.

Statute of the International Atomic Energy Agency (IAEA), 1957

The International Atomic Energy Agency is charged with making sure that nuclear materials, technology, and services that it provides be used only for peaceful purposes. A staff of inspectors was established to monitor all IAEA operations to ensure that they comply with health and safety as well as security and nonproliferation measures.

Antarctic Treaty, 1959

This treaty bans nuclear weapons, weapons tests, and disposal of nuclear wastes in Antarctica.

EURATOM Cooperation Act of 1958 (P.L. 85-846, 72 Stat. 1084, 42 USCA Sections 2291–2296). This law provided for a joint nuclear power program between the United States and the European Atomic Energy Commission (EURATOM), which included the governments of France, West Germany, Italy, Belgium, Netherlands, and Luxembourg. The program was to be carried out in accordance with provisions of the Atomic Energy Act of 1954 relating to international cooperation in atomic energy. The U.S. Atomic Energy Commission was authorized to enter into contracts for research and development and contracts with reactor operators to provide them with nuclear fuel (uranium or plutonium).

Strategically, this law sought to strengthen the economy and energy capabilities of U.S. allies in Europe, as part of bolstering defense against the threat of the Soviet Union, which would embark on its own program of nuclear cooperation with its client countries.

Treaty on the Non-Proliferation of Nuclear Weapons, 1968 (taking effect in 1970). This treaty, signed at London, Moscow, and Washington, D.C., prohibits signatory nations that possess nuclear weapons from transferring them to any other nation, or to aid any nonnuclear weapon state in producing such weapons. Signatory states not possessing nuclear weapons agree not to receive them from any other nation.

Nuclear Nonproliferation Act of 1978 (P.L. 95-242, 92 Stat. 120, 22 USCA 3201 et seq.). This law states that "the Congress finds and declares that the proliferation of nuclear explosive devices or the direct capability to manufacture or otherwise acquire such devices poses a grave threat to the security interests of the United States and to continued international progress toward world peace and development." This reflected the growing number of nations that had either demonstrated nuclear weapons capability prior to 1978 or were thought to be within a few years of being able to build nuclear weapons if they obtained fissionable material. There was also growing concern about the possibility that a terrorist group might be able to obtain or assemble a nuclear weapon. The NNPA therefore tightened international controls on the transfer of nuclear materials and in effect tied availability of reactor technology and fuel to a nation's willingness to cooperate in deploying systems and technologies that would minimize the risk of unauthorized access to nuclear materials. Nations that had not ratified the Treaty on the Non-Proliferation of Nuclear Weapons were encouraged to do so.

A nod to opposition to nuclear power was also included in instructing U.S. agencies to "cooperate with foreign nations in identifying and adapting suitable technologies for energy production and, in particular, to identify alternative options to nuclear power."

Regulatory Agencies

Although the Nuclear Regulatory Commission has overall responsibility for the licensing and regulation of nuclear power plants, a number of other federal agencies play a role in regulating activities in the nuclear industry.

Department of Defense

The Department of Defense has some regulations for use of and exposure to radioactive materials in the military.

Department of Energy

The Department of Energy establishes safety standards and regulations for its facilities, including the federal nuclear research laboratories and test facilities.

Department of Transportation

The Department of Transportation regulates transportation of radioactive materials in the United States. It administers the regulations in the Code of Federal Regulation (CFR) 49, parts 170–175, and has an Office of Hazardous Materials Safety that includes radioactive materials in its mission.

Environmental Protection Agency

The EPA issues regulations on the release of radioactive materials and gases into the environment (see CFR 40, parts 61, 141, 190, 192, 220–229, and 440). It also addresses radioactive contamination issues in its SuperFund cleanup program. The agency has a Radiation Protection Division (RPD) to administer these regulations.

Food and Drug Administration (FDA)

The FDA regulates the use of radiation, both nonionizing (such as lasers) and ionizing (radioisotope sources), in the production of food and pharmaceuticals and the use of medical devices employing radiation. It administers the regulation CFR 21, part 1000. It has a Center for Devices and Radiological Health.

Occupational Safety and Health Administration (OSHA)

OSHA is concerned with workplace safety in general. With regard to radiation safety, its regulations supplement those of the DOE and NRC. It administers CFR 29, parts 1910 and 1926.

Court Cases

Following are brief summaries of some important cases in federal courts involving nuclear power.

Baltimore Gas & Electric Co. v. NRDC, 462 U.S. 87 (1983)

The National Environmental Policy Act (NEPA) requires that all federal agencies consider the environmental impact of any federal action. The NRC decided, for purposes of the NEPA, that stored nuclear waste would not escape into the environment and therefore had no environmental impact. (This is called the "zero-release assumption.")

This rule was challenged as being arbitrary and capricious, since it failed to take into account the quite plausible possibility that stored wastes might leak into the environment. The Court ruled, however, that the NRC was reasonable in concluding that the chance of such releases was remote and that its consideration was sufficient to meet the requirements of the NEPA.

City of West Chicago v. Kerr-McGee, 677 F.2d 571 (7th Cir. 1982)

The City of West Chicago, Illinois, sued the Kerr-McGee Corporation, alleging that it had created a public nuisance at a nuclear plant by having open pits filled with chemical waste and refuse, as well as holes in floors and roofs. The defendant argued that the federal Atomic Energy Act preempted any state or local action against nuclear facilities. The 7th Circuit Court disagreed, saying that the City could regulate nonradiation-related hazards.

Commonwealth Edison Co. v. NRC, 819 F.2d 750–764 (7th Cir. 1987)

The Commonwealth Edison Company sued the Nuclear Regulatory Commission, arguing that the NRC should not be able to apply new 1984 inspection fee limits to work that had been done at its Byron and Braidwood (Illinois) plants before the new regulations took effect. After several procedural hearings, the court ruled that the NRC could apply its 1984 rules to the earlier work, and it assessed penalties and interest charges against Commonwealth Edison for the unpaid fees.

County of Suffolk v. Long Island Lighting Company, 728 F.2d 52 (2nd Cir. 1984)

In this case a county in New York tried to get a court order to allow an inspection of a nuclear power plant under construction in order to investigate claims of negligence, breach of contract, and misrepresentation and concealment in the design and construction of the plant. The court ruled that the county could not conduct the inspection because inspection of nuclear power plants falls within the areas of construction and operation of nuclear facilities, which are reserved to the Nuclear Regulatory Commission.

Critical Mass Energy Project v. NRC, No. 86–5647 (D.C. Cir. September 29, 1987, remanded in part, 644 F. Supp. 344 D.D.C. 1986)

The plaintiff, an antinuclear activist group, had sought access to documents of the Institute of Nuclear Power Operations (INPO) under the Freedom of Information Act (FOIA). The documents had been witheld under FOIA Exemption 4 as being "confidential."

The District of Columbia Circuit Court held that the documents in question were "commercial" and established a two-part legal test to determine whether they were also "confidential." The court agreed that the documents met the first part of the test and would not "customarily be released to the public." The INPO also had to show that release of the information would "impair [its] ability to obtain necessary information in the future." The court found the factual record inconclusive on this matter. However, the court narrowed its interpretation of the Freedom of Information Act in saying that impairment of other legitimate interests of the agency could also justify withholding the documents.

Duke Power Co. v. Carolina Env. Study Group, 438 U.S. 59 (1978)

In this case the Carolina Environmental Study Group, a labor union, and interested citizens asked the court to declare that the Price-Anderson Act, with its $560 million limitation on total liability for a nuclear accident, was unconstitutional. A federal district court agreed, ruling that the legislation was unconstitutional because the amount of liability payment it offered was far less than the potential damages that could be caused by the Duke power plant's operation (including radiation, pollution, and thermal pollution—excess heat). The court said this denied claimants the "due process" they were entitled to under the Fifth Amendment, and that it unfairly imposed a burden on potential victims of nuclear power instead of the burden being shared by the general public that benefited from having nuclear power.

Upon appeal by Duke, the U.S. Supreme Court ruled that the Price-Anderson limit did not deny due process rights under the Fifth Amendment and that the provision was "rationally related" to the intention of Congress to encourage the development of nuclear energy. The Court also pointed out that in the remote event of an accident whose damages exceeded the limit, Congress was free to provide special relief for the victims.

Eddleman v. NRC, 825 F.2d (4th Cir. 1987)

Wells Eddleman and several other persons filed a petition to review the NRC's licensing of the Shearon Harris nuclear power plant in North Carolina. The U.S. Circuit Court of Appeals for the Fourth Circuit unanimously ruled that the petitioners had no right to have a hearing before the decision to license was made. The court also stated that the petitioners had no right, under the Atomic Energy Act, to a hearing on a petition that had been considered and rejected by the commission as part of its immediate effectiveness review. The court also ruled that the NRC could grant the reactor license applicant an exemption from the requirement that a full-scale emergency planning exercise be held one year prior to licensing.

English v. General Electric Co., 496 U.S. 72 (1990)

Vera M. English, a technician employed in a GE nuclear plant, complained to management about safety problems, including the failure to clean up spills of radioactive materials. Frustrated by the lack of response, she refused to clean up such a spill herself, and instead outlined it in tape and pointed it out to management. GE ultimately fired her for failure to clean up the spill.

English filed a claim with the Secretary of Labor, charging that GE had violated a provision of the Energy Reorganization Act of 1974, which made it unlawful for an employer to retaliate against an employee for reporting safety violations. The hearing officer agreed GE's action violated the law but turned down English's claim because it had not been made in the required period of thirty days after the incident.

English then filed a suit in the U.S. District Court for the Eastern District of North Carolina that claimed, among other things, that GE had inflicted emotional distress on her. GE argued that such a state claim was preempted by the federal government because it related to radiation matters. The appeals court disagreed but nevertheless turned down English's claim as being in conflict with other provisions of the energy law.

The U.S. Supreme Court overturned the appeals court, ruling that English's claim of emotional distress was not preempted by the federal law. Among other reasons, it stated that the claim of emotional distress under state law was not closely related to the health and safety concerns preempted by the federal law, and that Congress had shown no intent to preempt such state claims.

Kerr-McGee v. Farley, 95-2121, U.S. (10 Cir. 1997)

In this case Kerr-McGee Corporation filed for a declaratory judgment to quash a suit brought against it in an Indian tribal court for injuries caused by uranium milling wastes. Kerr-McGee argued that under 1988 amendments to the Price-Anderson Act, any defendant in a tort (damages) action relating to radiation can require that the case be heard in federal court. The intention was to avoid having the nuclear industry deal with a large number of different jurisdictions.

The court, however, though accepting the intention of Price-Anderson, ruled that since the legislation did not explicitly deal with tribal courts—which enjoy considerable independence as agencies of sovereign tribes—Kerr-McGee could not force the tribal court to give up the case in favor of federal jurisdiction.

Metropolitan Edison v. People vs. Nuclear Energy, 460 U.S. 766 (1983)

In this case the NRC prepared to hold hearings as to whether the Three Mile Island 1 (TMI-1) plant could be reopened, TMI-2 having been subject to the famous disaster in 1979. The NRC invited citizens to testify about the desirability of reopening the plant and said they could include evidence of "indirect harm" including "psychological harm" caused by the trauma of the earlier accident. In actually holding the hearings, however, the NRC decided not to hear such indirect evidence. The antinuclear group People vs. Nuclear Energy filed a claim in the court of appeals arguing that the National Environmental Policy Act (NEPA) required that such evidence be considered. The appeals court agreed.

The Supreme Court, however, overturned the appeals court, ruling that NEPA required only that federal agencies consider harm to the environment itself, not possible indirect harm such as psychological distress.

New York v. United States, 488 U.S. 1041 (1992)

In this case (combined with two similar cases), New York challenged requirements of the Low-Level Radioactive Waste Act amendments of 1985. Among other things, this legislation gave states with nuclear waste facilities the right to charge increasingly high rates to waste producers in states that had no facilities and needed to dispose of waste, and even to cut off access to their facilities from states that failed to meet federal deadlines for establishing their own waste facilities. Further, if a state failed to provide for waste disposal after a date set by the federal government, the company generating the waste could force the state to take title (ownership) to the waste and thus responsibility for its disposal.

New York contended that among other arguments, these requirements violated the Tenth Amendment to the U.S. Constitution, which reserves to states or the people all powers not specifically granted to the federal government. The Court ruled, however, that though the federal government could not "commandeer" state officials and make them enforce federal requirements, Congress could, as part of its spending power, require that states perform certain actions in order to be eligible for federal funds—a common practice that in effect gives Congress powers that it would not otherwise have. Further, Congress could use its power to regulate interstate commerce to impose fees on the interstate transport of nuclear wastes. The Court did, however, strike down the provision that forced states to take title over wastes, saying that it was unconstitutionally coercive.

Northern Ind. Pub. Serv. Co. v. Walton League, 423 U.S. 12 (1975)

In this case the Supreme Court ruled that the Atomic Energy Commission's Licensing and Appeal Board was free to use a means of determining the population of the area around a nuclear plant that was different from that implied in its earlier regulations, because the method "sensibly conforms" with those regulations.

Ohio Citizens for Responsible Energy v. NRC, 803 F.2d 258 (6th Cir. 1986); Ohio v. NRC, 814 F.2d 258 (6th Cir. 1987)

These suits challenged the licensing of the Perry, Ohio, nuclear plant by the Nuclear Regulatory Commission. After an earthquake occurred ten miles from the plant, the petitioners requested that the NRC reopen the licensing and emergency planning procedures to consider seismic safety. The Sixth Circuit Court ruled that the NRC's decision not to reopen the licensing process was within its legitimate discretion.

Pacific Gas and Electric Company (PG&E) v. State Energy Resources Conservation and Development Commission, 103 U.S. 1713 (1983)

California had passed a statute requiring that a nuclear plant could be constructed only if it met federal requirements concerning disposal of high-level waste. PG&E argued that such regulation of nuclear plant construction was preempted by the NRC. However, the Supreme Court ruled that the state's regulation was concerned primarily with the economic impact of waste disposal, not the national security and public safety concerns that were the province of the NRC. Therefore the state's regulation was not preempted by the federal agency.

Roberts v. Florida Power and Light, 97-5195 (11 Cir. 1998)

This case arose when a suit was filed claiming that Bertram Roberts had developed leukemia as a result of exposure to radiation from FPL's Turkey Point nuclear power station. Federal legislation (including the Price-Anderson Act's 1988 amendments) specified that the procedures used by the federal court for determining liability would be "derived from" the laws of the state in which the injury took place, unless such procedures were inconsistent with federal legislation. The district court noted that since there had been no "extraordinary nuclear incident," Roberts would have to prove that the company had "breached its duty of care" to him by exposing him to a level of radiation that exceeded federal (not state) standards.

On appeal Roberts argued that since the legislation did not mention exposure levels, any level set by the state was not "inconsistent" with the federal legislation and should be used by the court in deciding his claim. The court, however, sided with FPL and the many precedents that gave the federal government the exclusive right to set standards for operation of nuclear plants.

Silkwood v. Kerr-McGee Corporation, 104 U.S. 615 (1984)

As part of a famous case that gained national attention, the State of Oklahoma had awarded punitive damages to Karen Silkwood for injuries caused by a leak at a federally licensed plutonium processing plant. The plant owner, Kerr-McGee, argued that the award should be set aside because it amounted to state regulation of radioactive hazards, an area preempted by the federal government. The company also argued that the award of damages was in conflict with public policy as embodied in the goal of the Atomic Energy Act of promoting nuclear power.

The Supreme Court rejected both arguments. It noted that the Atomic Energy Act's language about promoting nuclear power specifically stated that atomic energy be used only to the extent it was consistent with the safety and protection of the public.

State of Wisconsin v. Northern States Power Company, No. 85-CV-0032 (Cir. Ct. Dane Co., Wis., June 6, 1985)

In this case the circuit court ruled that the state could not regulate a utility's shipment of spent fuel within the state, such regulation being preempted by federal law.

Train v. Colorado Pub. Int. Research Group, 426 U.S. 1 (1976)

This case arose from a conflict between two sources of environmental regulations. On the one hand, the Federal Water Pollution Control Act (FWPCA) makes it unlawful to discharge "pollutants" into navigable waters without a permit from the Administrator of the Environmental Protection Agency and includes "radioactive materials" among the pollutants. On the other hand, the Atomic Energy Commission (and its successor, the Nuclear Regulatory Commission) under federal law have exclusive power to regulate discharges from nuclear power plants. When the EPA administrator refused to become involved in incidents of radioactive waste discharge, the Colorado Public Interest Research Group sued to force the EPA to intervene.

The district court dismissed the suit, saying that the AEC had sole regulatory power in such cases. The court of appeals, however, reversed that decision, noting in particular that the FWPCA specifically included radioactive materials.

The Supreme Court, however, ruled that the appeals court had erred in not considering the intent of Congress in drafting the FWPCA, which showed evidence that Congress had not intended to give the EPA any ability to regulate discharges by licensees subject to Atomic Energy Commission regulation.

Union of Concerned Scientists v. NRC, 824 F.2d 108 (D.C. Cir. 1987)

The Union of Concerned Scientists filed this suit to overturn the NRC's "backfitting" rule that allowed it to consider costs to the utility in establishing the "adequate protection standard." The court agreed and overturned the rule, stating that standards of adequate protection must be based solely on health and safety considerations, not cost. However, the court ruled that once this standard had been met, the NRC could take cost into account in deciding whether additional safety measures should be required.

United States of America and Trustees of Columbia University v. City of New York, 463 F. Supp. 604 (S.D.N.Y. 1978)

In this case the city had imposed a licensing requirement on a research and teaching reactor run by Columbia University. The court ruled that because the licensing requirements pertained to health and safety, federal law preempted the city from imposing them.

Vermont Yankee Power Corp. v. NRDC, 435 U.S. 519 (1978)

The National Resources Defense Council, an environmental group, challenged the adequacy of Licensing Board hearings regarding the Vermont Yankee nuclear plant. The NRDC charged that the NRC's hearings in which it granted Vermont Yankee's license had covered the transportation of nuclear fuel but not its processing or storage. When this decision was appealed to the NRC, it ruled that the processing and storage aspects were not highly significant. It did not allow cross-examination of witnesses at this hearing. The NRDC appealed to the courts, claiming that the hearing was inadequate.

The federal appeals court agreed with the NRDC that the hearings had been inadequate, but the U.S. Supreme Court reversed this ruling. The Court said that in general, unless the prevailing statute explicitly required certain procedures, an agency was free to establish its rulemaking procedures without interference by the courts.

5

Directory of Organizations

Hundreds of organizations are involved in some way with nuclear power. This chapter presents a selection of agencies and organizations that have been chosen for their importance, distinct character, and accessibility through the Internet. (Most organizations now have a website that is well-stocked with news, background information, and links to related sites.) Types of organizations listed include regulatory agencies, industry groups, scientific and professional societies, and environmental and other antinuclear groups.

Abalone Alliance
2940 16th St. #310
San Francisco, CA 94103
Phone: (415) 861-0592
Fax: (415) 558-8135
Website: http://www.sfo.com/~rherried/welcome.htm

This activist group is dedicated to opposing the nuclear industry and nuclear waste disposal while promoting safe energy alternatives. Its Energy Net website provides a variety of background facts, news, and documents about nuclear power, particularly in California, and antinuclear organizations.

Alliance for Nuclear Accountability (ANA)
1801 18th St. NW, Ste. 9-2
Washington, DC 20009
Phone: (202) 833-4668
Fax: (202) 234-9536

E-mail: ananuclear@earthlink.net
Website: http://www.ananuclear.org/

This is a network of more than thirty regional and national peace, antinuclear activist, and environmental groups. It focuses on trying to close unsafe weapons sites, production facilities, and waste storage facilities and on curtailing nuclear activities in general.

American Academy of Health Physics
1313 Dolley Madison Blvd., Ste. 402
McLean, VA 22101
Phone: (703) 790-1745, ext. 25
Fax: (703) 790-2672
E-mail: aahp@BurkInc.com
Website: http://www.hps1.org/aahp/

This is an organization for health physicists who study radiation effects and evaluate radiation protection methods. It is a good source for people considering study or a career in this field.

American Environmental Health Studies Project
6328 Strawberry Plains Pike
Knoxville, TN 37914
Phone: (423) 689-6631
E-mail: jackie@mindspring.com

This radiation and health project focuses especially on workers and neighbors of the Oak Ridge nuclear facilities. The group does research and community outreach and also provides legal and health advice to radiation victims.

American Institute of Physics (AIP)
1 Physics Ellipse
College Park, MD 20740-3843
Phone: (301) 209-3000
E-mail: aipinfo@aip.org
Website: www.aip.org

This is an organization of scientific societies devoted to the advancement of physics, scientific cooperation, and public outreach. It organizes conferences and publishes periodicals and books (through the AIP press). Its Niels Bohr Library is an important source of archives for research into the development of nuclear fission, the atom bomb, and nuclear reactors.

American Nuclear Society
555 N. Kensington Ave.
LaGrange Park, IL 60526
Phone: (800) 323-3044
E-mail: nucleus@ans.org
Website: http://www.ans.org/

This is an international organization for advancing and coordinating the professions of nuclear science and engineering as well as promoting research and education. Its approximately thirteen thousand members represent both the academic and corporate worlds. The organization publishes *Nuclear News, Radwaste Magazine,* and several peer-reviewed scientific journals.

American Physical Society: Division of Nuclear Physics
Website: http://www.phy.anl.gov/dnp/

This site describes recent developments and research opportunities in nuclear physics and includes brochures, physics education materials, and material on the history of nuclear physics.

Atomic Energy Commission (France)
CEA
31–33 Rue de la Fédération
75752 Paris cedex 15
France
Phone: 33-1-40-56-10-00
Website: http://www.cea.fr/ang/html/accueil.htm

The French national atomic energy agency (Commissariat a l'Énergie Atomique, or CEA) is charged with developing expertise in all areas of nuclear energy through research and education. The website has information about nuclear research and developments in France.

Atomic Energy Council (Taiwan)
Website: http://www.aec.gov.tw/meco/engl/e2.htm

Founded as an organization to promote nuclear research and development in Taiwan, the Atomic Energy Council took on regulatory responsibilities as the country built its first nuclear power plants and applications of radioisotopes in medicine and industry. Its website is a good source for information about nuclear developments in Taiwan.

Atomic Reclamation and Conversion Project
4117 Terrace St.
Oakland, CA 94611-5125
Phone: (510) 595-9500
Fax: (510) 517-5272
E-mail: lnvlvo@igc.org

An activity of the nonprofit Tide Center, this project is involved with radiation health studies and advocacy. Its representatives wrote a letter to President Clinton urging cancellation of the nuclear-powered Cassini space probe.

Australian Nuclear Science and Technology Organization (ANSTO)
PMB 1
Menai NSW
2234 Australia
Phone: 61-02-9717-3111
Fax: 61-02-9543-5097
E-mail: enquiries@ansto.gov.au
Website: http://www.ansto.gov.au/

ANSTO is Australia's national government-funded nuclear research, development, and operation organization, providing advice and information to the nuclear industry. Its website includes news and discussion of Australian and worldwide nuclear developments.

Ban Waste Coalition (Ward Valley, CA)
Website: http://www.enviroweb.org/wardvalley/

This site has background and news about the movement by Native Americans and other U.S. citizens to stop federal plans to establish an "interim" nuclear waste facility in Ward Valley in California's East Mojave desert. Activists cite environmental risk, cultural factors (Native American traditions), and ecological reasons (endangered species) in their opposition to the facility.

Bellona Foundation
P.O. Box 2141 Grunerlokka
0505 Oslo

Norway
Telephone: 47-23-23-46-00
Fax: 47-22-38-38-62
E-mail: info@bellona.no
Website: www.bellona.no

This is an environmental activist organization that focuses on issues involving nuclear waste, nuclear industry, and operations in the former Soviet Union (including Russian navy and other military nuclear operations).

California Energy Commission
Media and Public Communications Office
1516 9th St., MS-29
Sacramento, CA 95814-5504
Phone: (916) 654-4989
Fax: (916) 654-4420
Website: http://www.energy.ca.gov/nuclear/index.html

This state agency provides a variety of fact sheets, position papers, and other information about the shipment (or proposed shipment) of nuclear waste in or through California. The state advocates on behalf of its citizens with federal authorities, opposing, for example, use of the Concord Naval Weapons Station as a port of entry for nuclear waste.

Canadian Nuclear Society
144 Front St. W, Ste. 475
Toronto, Ontario
M5J 2L7 Canada
Phone: (416) 977-7620
Fax: (416) 979-8356
E-mail: pantonz@cna.ca
Website: http://www.cns-snc.ca/

This organization is dedicated to the advancement of nuclear science and technology in Canada through professional development and public education. It is open to all interested persons.

Citizens Clearinghouse for Hazardous Wastes
P.O. Box 6806
119 Rowell Ct.
Falls Church, VA 22046
Phone: (703) 237-2249

Fax: (703) 237-8389
E-mail: cchw@essential.org

This organization helps coordinate actions on a variety of waste issues, including radioactive waste.

Committee for Nuclear Responsibility
P.O. Box 421993
San Francisco, CA 94142
Phone/fax: (415) 776-8299
Website: http://www.ratical.com/radiation/CNR/

Chaired by prominent nuclear critic Dr. John Gofman, this organization strives to provide "independent analyses of the health effects and sources of ionizing radiation." The organization's website includes works by Gofman and others and links to other resources.

Ecologia (U.S. contact)
1 Main St.
Harford, PA 18823
Phone: (570) 434-9588
Fax: (570) 434-9589
E-mail: ecologia@igc.apc.org
Website: http://www.e-tip.org/

This Russian-based organization also has a U.S. office, indicated above. It is focused on environmental issues in the former USSR and pays considerable attention to radiation and nuclear power–related issues. Its website, called E-tip (Environmental Technical Information Project), provides a variety of documents, including a guide to citizen response to nuclear power plant accidents.

Electric Power Research Institute
3412 Hillview Ave.
Palo Alto, CA 94304
Phone: (650) 855-2000
Website: http://www.epri.com

This industry organization promotes public understanding of the electric industry and policies that would increase the supply of affordable electric energy. Nuclear power is among the technologies it promotes, and its website includes a page of links relating to nuclear energy technology.

European Nuclear Society
Belpstrasse 23, P.O. Box 5032
CH-3001 Bern
Switzerland
Phone: 41-31-320-61-11
Fax: 41-320-68-45
E-mail: ens@to.aey.ch
Website: http://www.euronuclear.org

This is an organization of about twenty thousand European nuclear scientists, engineers, and others interested in advancing nuclear science and engineering through collaboration and research. The organization has public outreach to promote a positive public awareness of nuclear power.

Fusion Power Associates
2 Professional Dr., Ste. 249
Gaithersburg, MD 20879
Phone: (301) 258-0545
Fax: (301) 975-9869
E-mail: fpa@compuserve.com
Website: http://ourworld.compuserve.com/homepages/fpa/

This is a nonprofit organization with participants from the corporate and research sectors that provides information about current research into fusion power as well as educational materials.

Greenpeace (U.S. office)
1436 U St. NW
Washington, DC 20009
Phone: (800) 326-0959
Website: www.greenpeaceusa.org

Greenpeace is a well-known worldwide environmental organization that is often in the forefront in confronting governments or companies that it views as enemies of the environment. It has been active in a number of nuclear issues. Greenpeace teams have intervened physically to try to stop nuclear testing.

Greenpeace International
Keizersgracht 176
1016 DW Amsterdam
The Netherlands

Phone: 31-20-523-62-22
Fax: 31-20-523-62-00
Website: www.greenpeace.org

This is the international Greenpeace organization. It serves as an umbrella group. See the above Greenpeace U.S. office listing for the American national group.

Hanford Education Action League (HEAL)
1718 W. Broadway, Ste. 203
Spokane, WA 99201
Phone: (509) 326-3370
Fax: (509) 326-2932
E-mail: healtm@aol.com
Website: http://www.iea.com/~heal/

HEAL is an activist organization based primarily in the Pacific Northwest and focusing on the Hanford nuclear weapons production facilities, which the organization charges have extensively polluted the environment and which pose serious risks of fire or explosion due to unsafe storage conditions.

Health Physics Society
1313 Dolley Madison Blvd., Ste. 402
McLean, VA 22101
Phone: (703) 790-1745
Fax: (703) 790-2672
E-mail: hps@BurkInc.com
Website: http://www.hps.org

The society is an international organization for health physicists. It is dedicated to professional practice in occupational and environmental radiation safety and also provides public information.

Institute for Energy and Environmental Research
6935 Laurel Ave.
Takoma Park, MD 20912
E-mail: ieer@ieer.org
http://www.ieer.org/index.html

This organization seeks to bring quality science into the service of democratic decisionmaking about the environment. Nuclear-related issues it addresses include handling and release of radioactive materials, such as contamination of water by tritium.

Institute of Nuclear Power Operations
700 Galleria Pkwy. NW
Atlanta, GA 30339-5957
Phone: (770) 644-8000

This organization was established in 1979 following the Three Mile Island accident. Its membership includes all U.S. utilities that operate nuclear power plants. Among its activities are:

- Analysis of reported events and dissemination of the lessons learned
- Promoting the exchange of information and good practices among all nuclear utilities
- Benchmarking against international best practice
- Developing with industry, and monitoring, a set of ten performance indicators
- Maintaining evaluation and peer review programs

International Atomic Energy Agency (IAEA)
P.O. Box 100
A-1400 Vienna
Austria
Phone: 011-4-31-2060
Website: http://www.iaea.org
E-mail: Official.Mail@iaea.org

The IAEA is a United Nations agency that coordinates efforts of many nations' governments in the nuclear field. It has attempted to realize President Eisenhower's 1957 call for "Atoms for Peace."

International Nuclear Safety Center (INSC)
Argonne National Laboratory, Bldg. 208
Dave Hill, INSC Director
9700 S. Cass Ave.
Argonne, IL 60439
Phone: (630) 252-5167
Fax: (630) 252-6690
E-mail: inscdb@anl.gov
Website: http://www.insc.anl.gov

The INSC is an international organization of nuclear professionals and groups. It is a strongly pronuclear group that tries to improve the safety of nuclear installations throughout the world and to educate and improve the public perception of the risks and benefits of nuclear power. The website includes numerous links to reports.

Nuclear Control Institute (NCI)
1000 Connecticut Ave. NW, Ste. 804
Washington, DC 20036
Phone: (202) 822-8444
Fax: (202) 452-0892
E-mail: nci@mailback.com
Website: http://www.nci.org/home.htm

The NCI is a research and advocacy organization working to prevent nuclear proliferation. It seeks to mobilize political pressure to enforce nonproliferation agreements. The institute's website provides a variety of fact sheets, studies, and news releases.

Nuclear Energy Agency (U.S. office)
The OECD Washington Centre
2001 L St. NW, Ste. 650
Washington, DC 20036-4922
Phone: (202) 785-6323
Fax: (202) 785-0350
E-mail: neapub@nea.fr
Website: http://www.nea.fr/welcome.html

This is an international organization within the Organization for Economic Cooperation and Development (OECD). Its purpose is to promote the development of nuclear energy as an environmentally safe and economical energy source. It operates through the cooperation of its twenty-seven participating governments, including countries in Europe, North America, and Asia. Efforts include promotion and maintenance of effective nuclear safety programs, monitoring exposure incidents, coordinating emergency response plans, and developing effective waste management systems.

Nuclear Energy Institute
1776 I St. NW, Ste. 400
Washington, DC 20006-3708
Phone: (202) 739-8000 (daytime); (703) 644-8805 (evenings)
E-mail: media@nei.org
Website: http://www.nei.org

The Nuclear Energy Institute is the principal public education and lobbying organization of the nuclear industry, presenting a variety of facts about nuclear power from a resolutely pronuclear viewpoint. Primarily based in the American nuclear industry, the institute also has about three hundred organizational members worldwide.

Nuclear Information and Resource Service (NIRS)
1424 16th St. NW, Ste. 601
Washington, DC 20036
Phone: (202) 328-0002
Fax: (202) 462-2183
E-mail: nirnet@nirs.org
Website: http://www.nirs.org/

NIRS describes itself as "the information and networking center for citizens and environmental organizations concerned about nuclear power, radioactive waste, radiation, and sustainable energy issues." It conducts a variety of campaigns, such as opposition to creation of "interim" waste dumps, opposition to fuel reprocessing, regional "nuclear free" campaigns, and cooperation with activists in Eastern Europe.

Nuclear Waste Citizens Coalition
c/o Citizen Alert
P.O. Box 17173
Las Vegas, NV 89114
Phone: (702) 796-5662
Fax: (202) 796-4886
E-mail: citizenalert@igc.org
Website: http://www.igc.org/citizenalert/nwcc/

This is a coalition of eighteen regional and national organizations involved with a broad range of antinuclear activities, including opposition to nuclear reactors, waste sites, and weapons. The website includes fact sheets and links.

Public Citizen: Critical Mass Energy Project
1600 20th St. NW
Washington, DC 20009
E-mail: cmep@citizen.org
Website: http://www.citizen.org/CMEP/

This project, part of Ralph Nader's Public Citizen organization, includes among its activities campaigns relating to nuclear safety, nuclear waste, radioactive recycling (fuel reprocessing), and food irradiation.

Radiation Effects Research Foundation
Hiroshima Laboratory
5-2 Hijiyama Park, Minami-ku
Hiroshima City, 732-0815
Japan
Phone: 81-82-261-3131
Website: http://www.rerf.or.jp/eigo/experhp/rerfhome.htm

The foundation is a cooperative U.S.-Japanese site organization that focuses on the effects of radiation on human health, with emphasis on survivors of the atomic bombings of World War II but with general applicability to other radiation hazards. Its website includes an introduction to radiation effects, answers to frequently asked questions, and summaries of studies carried out by the laboratories at Hiroshima and Nagasaki.

Radioactive Waste Management Associates
526 W. 26th St., Rm. 517
New York, NY 10001
Phone: (212) 620-0526
Fax: (212) 620-0518
E-mail: radwaste@rwma.com
Website: http://www.igc.apc.org/RWMA/

This is a consulting firm specializing in aiding governments and citizen organizations in dealing with radioactive waste management issues as well as testifying on behalf of plaintiffs in personal injury and property damage cases. Its website provides links to government and private sites as well as documents and news about waste issues.

STAR (Standing for Truth About Radiation)
STAR Foundation
66 Newtown Ln., Ste. 3
P.O. Box 4206
East Hampton, NY 11937
Phone: (516) 324-0655
Fax: (516) 324-2203

E-mail: info@noradiation.org
Website: http://www.noradiation.org/main.htm

STAR is an organization founded by Long Island antinuclear activists concerned about radiation dangers at the Millstone nuclear power station and the Brookhaven National Laboratory. The organization seeks to hold such institutions accountable for their effects on health and the environment and to educate and mobilize the public.

Union of Concerned Scientists
2 Brattle Square
Cambridge, MA 02238-9105
Phone: (617) 547-5552
Fax: (617) 864-9405
E-mail: ucsoutreach@ipc.apc.org
Website: http://www.ucsusa.org/

This is an organization of scientists concerned with global environmental, energy policy, and technology issues. Nuclear power safety is a major focus of the organization's activities. The organization produces publications including a curriculum guide for grades seven through twelve.

United States Department of Energy
1000 Independence Ave. SW
Washington, DC 20585
Phone: (202) 586-5000
Website: http://www.doe.gov

The Department of Energy plays a major role in funding energy research and in promoting energy efficiency and alternative forms of power.

United States Department of Energy: Energy Information Administration
Phone: (202) 586-8800
E-mail: infoctr@eia.doe.gov
Website: http://www.eia.doe.gov/

This DOE office is tasked with developing statistical data about energy sources and the energy market. It also provides public access to a wide variety of information, including recent developments and economic forecasts. Of interest to researchers in nuclear energy is the annual summary of the uranium industry available at the website. The FAQ file at the website provides information about the scope of information available and how to access it.

United States Department of Energy: Office of International Nuclear Safety and Cooperation
19901 Germantown Rd.
Germantown, MD 20874
Phone: (301) 903-0234
Fax: (301) 903-4211
E-mail: richard.reister@hq.doe.gov
Website: http://atom.pnl.gov:2080/

The DOE's Office of International Nuclear Safety and Cooperation is a comprehensive program with U.S. experts and resources seeking to help improve safety at the Soviet-designed nuclear power plants in Russia and Eastern European countries. Its website includes technical and historical background on these facilities, a chronology of events, and other resources.

United States Department of Energy: Office of Science
Website: http://www.er.doe.gov/

This DOE office is in charge of funding basic research programs in energy technology and in the relationship between energy and the environment. Research areas include physics, fusion, advanced scientific computing, and biological/environmental aspects of energy. The agency builds and manages a variety of research facilities, both in national laboratories and in conjunction with universities. The agency's website offers a variety of news stories about research programs, a newsletter called *DOE Pulse* (available on-line), and background information and links.

United States Department of Energy: Office of Scientific and Technical Information
Website: http://www.osti.gov/

This office is responsible for distributing information generated by DOE research projects. Its website provides access to publications including the DOE Information Bridge, a full-text database.

United States Environmental Protection Agency: Radiation Protection Program
Ariel Rios Bldg.
1200 Pennsylvania Ave. NW

Washington, DC 20460
Phone: (202) 260-2090
Website: http://www.epa.gov/radiation/

The Environmental Protection Agency has a variety of responsibilities with regard to radiation hazards, including developing standards for radioactive materials in the environment, assessing risks and technologies, and developing emergency response programs.

United States Nuclear Regulatory Commission
1717 H St. NW
Washington, DC 20555
Phone: (202) 634-1481
Website: http://www.nrc.gov

Founded in 1974, this is the federal agency responsible for the protection of the public health and safety in the use of nuclear materials. It regulates reactor operation, fuel cycle operations, and the transport, storage, and disposal of nuclear materials and waste.

Uranium Information Centre (Australia)
GPO Box 1649N
Melbourne 3001
Australia
Phone: 61-03-9629-7744
Fax: 61-03-9629-7207
E-mail: uic@mpx.com.au
Website: http://www.uic.com.au

This organization provides extensive information about the Australian uranium industry and nuclear power production. The website includes news, briefing papers, and resource links.

The Uranium Institute
12th Fl., Bowater House
114 Knightsbridge
London SW1X 7LJ
United Kingdom
Phone: 44-1-71-225-0303
Fax: 44-1-71-225-0308
E-mail: ui@uilondon.org
Website: http://www.uilondon.org/index.htm

The Uranium Institute is an international organization that provides information about nuclear energy. It is a strong advocate of nuclear power. Topics on its website include the nuclear fuel cycle, safety issues, and the role of nuclear power in forestalling global warming.

Western States Legal Foundation
1440 Broadway, Ste. 500
Oakland, CA 94612
Phone: (510) 839-5877
Fax: (510) 839-5397
E-mail: wslf@igc.apc.org

This activist group focuses on abolishing nuclear weapons but also devotes effort to the hazards of nuclear power and nuclear waste. It organizes and contributes to legal actions to further its goals.

WISE (World Information Service on Energy)
Wise Amsterdam
P.O. Box 59636
1040 LC Amsterdam
The Netherlands
Phone: 31-20-612-63-68
Fax: 31-20-689-21-79
E-mail: wiseamster@antenna.nl
Website: http://www.antenna.nl/wise/

This Amsterdam-based grassroots antinuclear organization maintains a website with news about nuclear developments, risks, and hazards as well as protest actions. It issues a regular newsletter that is available by subscription or, soon, on the Net.

World Association of Nuclear Operators
U.S. Regional Office
700 Galleria Pkwy. NW
Atlanta, GA 30339-5957
Phone: (770) 644-8000
Website: http://www.insc.anl.gov/neisb/neisb4/NEISB_1.4.html

This organization was founded in response to public concern following the nuclear accident at Chernobyl. It seeks to strengthen the self-regulation efforts of the nuclear industry, develop sound safety practices, provide a mechanism for prompt reporting of nuclear incidents, and restore public trust in the industry.

6

Print Resources

An explosion in print publications on nuclear power issues occurred during the 1970s and early 1980s, peaking just after the Three Mile Island accident. Many of the arguments made in the "pro" and "con" nuclear books of the time are still relevant today, even though many of the facts used to support them are outdated.

The mid-1980s saw a peak of interest in nuclear disarmament and the possibility of nuclear war, which is outside the scope of this book. Although the Chernobyl disaster of 1986 again put nuclear power in the headlines, the issue seemed to remain dormant.

However, the 1990s saw something of a revival of writing on nuclear power issues. Areas of renewed concern included the seemingly intractable problem of disposing of the growing piles of nuclear waste, and the threat of nuclear proliferation and terrorism following the breakup of the Soviet Union.

Selected books from the earlier era that have lasting historical or philosophical interest are included in the following list, but the focus is on publications of the 1990s, which offer the most current facts and analyses.

The books listed in this chapter are divided into the following general topics and subtopics:

References and Overviews
- Bibliographies
- General reference and overviews (including balanced "pro and con" surveys)

Advocacy
- Antinuclear advocacy, based on broad issues
- Pronuclear advocacy

Regulations, Politics, and Policymaking
- Regulatory policy and politics, often from a historical perspective
- Regulatory case studies (such as those dealing with a specific nuclear plant)
- Future prospects for regulation or policymaking

International Issues
- Nuclear developments in other nations
- Nuclear proliferation issues

Nuclear Disasters
- Three Mile Island
- Chernobyl

Nuclear Waste Issues
- Environmental effects of nuclear waste
- Waste disposal and management issues

Miscellaneous Issues
- Fusion power
- Nuclear power in space
- Other

Periodicals

References and Overviews

Bibliographies

Atomic Energy Commission Group, United Nations Department of Security Council Affairs. ***An International Bibliography on Atomic Energy.*** New York: United Nations, 1949–1952. 2 volumes and supplements.

This is an exhaustive bibliography of works from the earliest years of the development of nuclear energy. Covers books, articles and papers, pamphlets, and audiovisual materials.

Babkina, A. M. ***Nuclear Proliferation: An Annotated Bibliography.*** Huntington, N.Y.: Nova Science Publishers, 1999. 245 pages. ISBN 1560726466.

This is an extensive annotated bibliography on nuclear weapons proliferation issues. Its 643 entries are divided into two main sections: books 1985–1999 and journal articles 1994–1999.

Burns, Grant. ***The Atomic Papers: A Citizen's Guide to Selected Books and Articles on the Bomb, the Arms Race, Nuclear Power, the Peace Movement, and Related Issues.*** Metuchen, N.J.: Scarecrow Press, 1984. 309 pages. ISBN 081081692X.

This is an out-of-print but useful bibliography of books and articles related to the development of (and opposition to) nuclear policies, including nuclear power issues. It is subdivided by subject. The author is unabashedly antinuclear, but selections are comprehensive.

Casper, Dale E. ***Protest and Public Policy: World-Wide Anti Nuclear Movement, 1982–1986.*** Monticello, Ill.: Vance Bibliographies, 1987. 11 pages. ISBN 1555904068.

This is a brief bibliography, but it is useful because of its international focus on the antinuclear movement of the early to mid-1980s.

Department of Energy. Technical Information Center Staff. ***Radioactive Waste Management: Airborne Radioactive Effluents: Releases and Processing: A Bibliography.*** U.S. Dept. of Energy, 1982. 245 pages. ISBN 0870794809.

Department of Energy. Technical Information Center Staff. ***Radioactive Waste Management: Decontamination and Decommissioning: A Bibliography.*** U.S. Dept. of Energy, 1982. 124 pages. ISBN 087079485X.

Department of Energy. Technical Information Center Staff. ***Radioactive Waste Management: Formerly Utilized Sites—Remedial Action: A Bibliography.*** U.S. Dept. of Energy, 1982. 47 pages. ISBN 0870794868.

Department of Energy. Technical Information Center Staff. ***Radioactive Waste Management: High-Level Radioactive Wastes: A Bibliography.*** Lynda H. McLaren, ed. U.S. Dept. of Energy, 1984. 393 pages. ISBN 0870795295.

Department of Energy. Technical Information Center Staff. ***Radioactive Waste Management: Low-Level Radioactive Waste: A Bibliography.*** U.S. Dept. of Energy, 1984. 184 pages. ISBN 0870795252.

Department of Energy. Technical Information Center Staff. ***Radioactive Waste Management: Nuclear Fuel Cycle: A Bibliography.*** U.S. Dept. of Energy, 1984. 138 pages. ISBN 0870795333.

Department of Energy. Technical Information Center Staff. ***Radioactive Waste Management: Nuclear Fuel Cycle Reprocessing: A Bibliography.*** U.S. Dept. of Energy, 1982. 248 pages. ISBN 0870795074.

Department of Energy. Technical Information Center Staff. ***Radioactive Waste Management: Spent Fuel Storage: A Bibliography.*** U.S. Dept. of Energy, 1982. 154 pages. ISBN 0870794787 (a supplement was published in 1984, ISBN 087079535X).

Department of Energy. Technical Information Center Staff. ***Radioactive Waste Management: Transuranic Wastes: A Bibliography.*** U.S. Dept. of Energy, 1982. 147 pages. ISBN 0870794825 (a supplement was published in 1985, ISBN 0870795724).

Department of Energy. Technical Information Center Staff. ***Radioactive Waste Management: Uranium Mill Tailings: A Bibliography.*** U.S. Dept. of Energy, 1982. 106 pages. ISBN 0870794930 (a supplement was published in 1985, ISBN 0870795740).

Department of Energy. Technical Information Center Staff. ***Radioactive Waste Management: Waste Isolation: A Bibliography.*** U.S. Dept. of Energy, 1982. 297 pages. ISBN 0870795058 (a supplement was published in 1985, ISBN 0870795767).

Frankena, Frederick. ***Ethics and Values in Radioactive Waste Management and Disposal: a Bibliography.*** Monticello, Ill.: Vance Bibliographies, 1991. 7 pages. ISBN 0792007514.

This is a brief bibliography of considerations of the "big picture" behind the nuclear waste controversy.

Frankena, Frederick, and Joanne Koelln Frankena. ***Citizen Participation in Nuclear Power Decision Making: A Bibliography.*** Monticello, Ill.: Vance Bibliographies, 1987. 13 pages. ISBN 1555904432.

This brief bibliography is useful because of its narrow focus on an important aspect of the antinuclear movement.

Frankena, Frederick, and Joanne Koelln Frankena. ***Implications of the Chernobyl Nuclear Accident for Policy and Planning: A Bibliography.*** Thousand Oaks, Calif.: Sage Publications, 1992. n.p. ISBN 0866022848.

This is a brief bibliography that captures the important literature arising from the worldwide reaction during the five years following the Chernobyl accident.

Frankena, Frederick, and Joanne Koelln Frankena. ***Radioactive Mining and Milling Waste: A Bibliography of Policy, Politics, and Law.*** Monticello, Ill.: Vance Bibliographies, 1991. 17 pages. ISBN 0792007530.

Frankena, Frederick, and Joanne Koelln Frankena. ***Radioactive Waste as a Social and Political Issue: A Bibliography.*** New York: AMS Press, 1991. 668 pages. ISBN 0404616283.

Frankena, Frederick, and Joanne Koelln Frankena. ***Radioactive Waste Problems at Commercial Nuclear Facilities and Disposal Sites: A Bibliography.*** Monticello, Ill.: Vance Bibliographies, 1991. 13 pages. ISBN 0792007557.

Gabriel, Michael R. ***Nuclear Energy and Public Safety.*** Chicago: CPL Bibliographies, 1982. 2 vols. ISBN 086602073X (v. 1); 0866020748 (v. 2).

The first volume is a bibliography of technical literature on nuclear power, and the second volume covers popular literature.

Hassler, Peggy M. ***Three Mile Island: A Reader's Guide to Selected Publications and Government-Sponsored Research Publications.*** Metuchen, N.J.: Scarecrow Press, 1988. 214 pages. ISBN 0810821184.

This is a research guide and annotated bibliography covering many of the important sources for scholars investigating the Three Mile Island nuclear accident and its aftermath. It includes government reports and research studies.

International Atomic Energy Agency. ***International Atomic Energy Agency Publications, 1980–1992.*** Vienna, Austria: Division of Publications, International Atomic Energy Agency, 1993. 6 volumes.

This is a catalog and bibliography of the agency's publications. Topic areas include earth sciences, life sciences, nuclear measurements, techniques and instrumentation; industrial applications; plasma physics and nuclear fusion; nuclear power; nuclear fuel cycle; nuclear techniques in food and agriculture; and nuclear and radiological safety.

Klema, Ernst D., and Robert L. West. ***Public Regulation of Site Selection for Nuclear Power Plants: Present Procedures and Reform Proposals—An Annotated Bibliography.*** Washington, D.C.: Resources for the Future, 1977. 151 pages. ISBN 0608134937.

This is an extensive bibliography for specialized research in this topic. Generally this must be ordered through a bookstore or from Books on Demand on-line.

Nuclear America: A Historical Bibliography. Santa Barbara, Calif.: ABC-CLIO Information Services, 1984. 184 pages. ISBN 0874363608.

This bibliography contains abstracts of 824 articles on nuclear power in the United States drawn from the ABC-CLIO Information Services Database. Entries are organized into the following topics: "The Road to Hiroshima," "The Development of Nuclear Energy," "The Balance of Terror," "Attempts at Nuclear Arms Control," and "Nuclear Reactors and Public Reaction."

Vance, Mary A. ***Nuclear Hazards Insurance: A Bibliography.*** Monticello, Ill.: Vance Bibliographies, 1989. 13 pages. ISBN 0792002199.

This is another specialized bibliography. The inadequacy of insurance or government guarantees in case of nuclear disaster is an important argument made by opponents of nuclear power.

Vance, Mary A. ***Nuclear Nonproliferation: A Bibliography.*** Monticello, Ill.: Vance Bibliographies, 1989. 20 pages. ISBN 0792003535.

This bibliography includes works about nonproliferation negotiations and issues up to the end of the Soviet era.

Voress, Hugh E. ***Bibliographies of Interest to the Atomic Energy Program.*** Rev. ed. Washington, D.C.: U.S. Atomic Energy Commission, 1958. (TID-3043).

This is a collection of bibliographies covering development of the nuclear energy program to the mid-1950s.

Wood, M. Sandra, and Suzanne M. Shultz. ***Three Mile Island: A Selectively Annotated Bibliography.*** New York: Greenwood Press, 1988. 309 pages. ISBN 0313255733.

This is a bibliography that compiles references to literature on all aspects of the Three Mile Island nuclear accident and its aftermath. Entries are drawn from science (medicine, nuclear biology, and the environment), social studies, and the humanities. The introductory essay by Pennsylvania Lieutenant Governor William W. Scranton III provides a good overview of both the events surrounding the accident itself and the reactions of public opinion, scientists, and government agencies. Entries include books, monographs, journal articles, reports, hearing transcripts, and other government documents.

General Reference and Overview

Ackland, Len, and Stephen McGuire. ***Assessing the Nuclear Age: Selections from the Bulletin of the Atomic Scientists.*** Chicago: Educational Foundation for Nuclear Science, 1986. 382 pages. ISBN 0941682080.

The *Bulletin of the Atomic Scientists* has been an influential voice in issues involving nuclear weapons and energy policy. This collection is a useful and varied presentation of articles on nuclear issues. (Although discussion of nuclear weapons and related policies predominates, there is also useful discussion of energy and power issues, and of course the weapons and power arenas are often related.)

Atkins, Stephen E. ***Historical Encyclopedia of Atomic Energy.*** Westport, Conn.: Greenwood Publishing, 2000. 504 pages. ISBN 0313304009.

This comprehensive encyclopedia covers the development of nuclear energy and weapons in more than 450 entries. Written for readers of high-school level to general adult level. Includes entries for scientific concepts and processes, nuclear accidents and protest groups, legal and governmental actions, and biographies of key individuals. Also includes a timeline and recommended readings.

Bodansky, David. ***Nuclear Energy: Principles, Practices, and Prospects.*** Woodbury, N.Y.: American Institute of Physics, 1996. 396 pages. ISBN 1563962446.

This is a fairly technical review of nuclear power development and the theory and practice underlying nuclear power generation. Topics include elements of nuclear reactions, the nuclear fuel cycle, waste disposal, and the operational and economic considerations for nuclear plants. Accessible to readers who have had basic college physics and some math, though a physics primer in the appendices can help refresh the reader's grasp of these topics.

Cassedy, Edward S., and Peter Z. Grossman. ***Introduction to Energy: Resources, Technology and Society.*** 2d ed. New York: Cambridge University Press, 1998. 427 pages. ISBN 0521637678.

This is an excellent introduction to energy alternatives for the general reader. Discusses the overall issues relating to electricity generation and compares the economics and environmental impacts of both conventional energy sources (including nuclear fission) and alternatives such as solar cells, alcohol-based fuels, wind, solar power, and nuclear fusion. Includes appendices that explain underlying scientific principles such as basic thermodynamics.

Howes, Ruth, and Anthony Fainberg, eds. ***The Energy Source Book: A Guide to Technology, Resources and Policy.*** New York: American Institute of Physics, 1991. 536 pages. ISBN 088318706X.

This is an excellent though somewhat technical collection of papers on a variety of energy sources, including fossil fuels, nuclear fission and fusion, solar, hydroelectric, geothermal, wind, and biomass. Includes evaluations both of the feasibility of the various technologies and of their application in various areas such as transportation, manufacturing, and agriculture.

International Atomic Energy Agency. ***Nuclear Power Reactors in the World.*** 1982– . ISSN 10112642.

This is a serial publication giving facts and statistics about nuclear power reactors, compiled by the International Atomic Energy Agency.

Kaku, Michio, and Jennifer Trainer, eds. ***Nuclear Power—Both Sides: The Best Arguments for and against the Most Controversial Technology.*** New York: Norton, 1982. 279 pages. ISBN 0393016315.

This collection contains twenty-one essays for and against nuclear power by many of the most prominent individuals involved in the controversy during the 1970s. Contributors include Ralph Nader, John Gofman, Hans Bethe, and Bernard Cohen. Although the editors are antinuclear, the selection and presentation of the material are generally well balanced. Since the core arguments have changed little in the past twenty years, the book remains a valuable background resource for the nuclear controversy.

Nuclear Power Plants Worldwide. Detroit: GALE Research Inc., 1993– . ISSN 10679979.

This extensive reference work provides both a directory giving information about nuclear power plants around the world and an introduction and history of nuclear power development. Information for each plant includes its current status, contact information for plant management, general technical specifications (type of reactor, electrical output capacity, etc.), chronology of its construction, and economic factors (including operating costs). The coverage also includes a profile of the nuclear industry in each country and an index of types of technical problems or accidents with a list of plants that have suffered each situation. The publication is currently issued as a serial approximately every three years.

Porro, Jeffrey, ed. ***The Nuclear Age Reader.*** New York: Knopf, 1989. 541 pages. ISBN 0394382617.

A companion to the PBS series *War and Peace in the Nuclear Age,* this extensive anthology includes government documents, critical essays and articles, and other materials from the 1940s through the 1980s.

Warf, James C. ***All Things Nuclear.*** Los Angeles: Southern California Federation of Scientists, 1989. ISBN 096267060X.

This is a good overview for placing nuclear power in the broad context of nuclear issues, including weapons, nuclear policy, and nuclear diplomacy.

Wolfson, Richard. ***Nuclear Choices: A Citizen's Guide to Nuclear Technology.*** Rev. ed. Cambridge, Mass.: MIT Press, 1993. 467 pages. ISBN 0262731088.

This book provides a comprehensive and clearly written introduction to nuclear issues. It begins with fundamentals of the atom and radioactivity, then introduces nuclear power in the context of energy alternatives and discusses the disposal of nuclear wastes. The final section deals with nuclear weapons. Takes a generally pro–nuclear power viewpoint but includes many examples of accidents and other problems.

Advocacy

Antinuclear

Brown, Jerry, and Rinaldo Brutoco. ***Profiles in Power: The Antinuclear Movement and the Dawn of the Solar Age.*** New York: Twayne Publishers, 1997. 249 pages. ISBN 0805738797.

This collection profiles the stories of ten persons from varying backgrounds (such as a biologist, grassroots activist, and a school teacher) who became outspoken opponents of the American nuclear industry. The stories are presented in the context of the study of social movements and the implications of the rise of the grassroots antinuclear (and prosolar) movements for social change in a democracy. (Note: The first author is not the former California governor.)

Caldicott, Helen. ***Nuclear Madness: What You Can Do.*** Rev. ed. New York: Norton, 1994. 240 pages. ISBN 0393310116.

This is a revised and expanded edition of a critique of nuclear power by one of the best-known antinuclear activists. Caldicott views "nuclear madness" as part of a pervasive influence of a powerful nuclear industry that ties in to destructive attitudes in our culture. She believes that the end of the Cold War and the reduction in the growth of the nuclear industry should not lead to complacency. The continued growth of nuclear power in the Third World and the problem of a growing stockpile of dangerous nuclear wastes continue to pose critical issues. She personally challenges the reader to come to terms with the moral issues and fight for a nuclear-free future.

Garcia-Gorena, Velma. ***Mothers and the Mexican Antinuclear Power Movement.*** Tucson: University of Arizona Press, 1999. 187 pages. ISBN 0816518750.

This is a sociological study of the development of the antinuclear movement in Mexico and the personalities who played a role in its struggles. The movement, largely organized by women ("Madres Veracruzanas"), arose when the Mexican government began to build the Laguna Verde nuclear power plant. The study concludes with a feminist analysis of the politics and fate of the movement. Includes a chronology and bibliography.

Gofman, John William. ***Poisoned Power: The Case against Nuclear Power Plants before and after Three Mile Island.*** Emmaus, Pa.: Rodale Press, 1979. 353 pages. ISBN 0878572880.

This classic antinuclear work by a physicist-activist sees Three Mile Island as the culmination and epitome of irresponsible policies and procedures that the authors believe are inherently risky and destructive. Further, the author insists that nuclear power is unnecessary for providing a sustainable future with adequate energy resources. Full text is available on-line at http://www.ratical.com/radiation/CNR/PP/.

May, John. ***The Greenpeace Book of the Nuclear Age: The Hidden History, the Human Cost.*** New York: Pantheon Books, 1989. 378 pages. ISBN 0679729631.

This is a comprehensive indictment of nuclear weapons testing and processing facilities for their effects on workers, the general public, and the environment. Discusses accidents and their consequences.

Sims, Gordon H. E. ***The Anti-Nuclear Game.*** Ottawa: University of Ottawa Press, 1990. 285 pages. ISBN 0776602853.

The author offers an exposé of the antinuclear movement and a spirited defense of nuclear power. Sims describes his years as an environmental activist and nuclear critic during the 1970s, but then recounts how his views changed as he considered the realistic risks and many advantages of nuclear power for meeting the energy needs of the twenty-first century. He claims that nuclear critics go out of their way to mislead and frighten the public.

Pronuclear

Beckmann, Petr. ***The Health Hazards of Not Going Nuclear.*** Boulder, Colo.: Golem Press, 1976. 190 pages. ISBN 0911762167.

This is an older but still very cogent defense of nuclear power. The author looks at the underpublicized hazards of the most prevalent kinds of nonnuclear power (particularly coal) and argues that not building nuclear plants to meet growing energy demand will cause health risks because more hazardous conventional plants will be built instead. (In the years since the book was written, improved energy efficiency has reduced energy demand projections somewhat. Nuclear opponents generally urge solar or other environmentally benign sources of power as alternatives for meeting energy demands.)

Cohen, Bernard L. ***The Nuclear Energy Option: An Alternative for the 90s.*** New York: Plenum Press, 1990. 338 pages. ISBN 0306435675.

The author makes a forceful case for nuclear power, dispelling what he considers to be misconceptions in the public mind. He emphasizes benefits and argues that safety risks and waste issues can be addressed by modern technology, including a new generation of safer reactors. Besides being safer than ever before, nuclear power can provide energy independence and reduces use of fossil fuels that endanger both health and the environment.

Lilienthal, David E. ***Atomic Energy: A New Start.*** New York: Harper and Row, 124 pages. 1980. ISBN 0060126175.

The author, the first chairman of the Atomic Energy Commission, describes his experiences and draws conclusions about nuclear issues. He supports nuclear energy in general but expresses safety concerns and discusses alternatives (such as new reactor designs) that may make the industry safer.

Waltar, Alan E. ***America the Powerless: Facing Our Nuclear Energy Dilemma***. Madison, Wis.: Cogito Books, 1995. 235 pages. ISBN 0944838588.

This book is written for a popular audience. It advocates the expansion of nuclear power as the only alternative to the United States becoming dependent on oil-producing countries. It also argues that nuclear power is much better for the environment than fossil fuels, since it neither produces air pollution nor contributes to the greenhouse effect that has become a major concern. The author is a past president of the American Nuclear Society.

Regulations, Politics, and Policymaking

Regulatory Policy and Politics

Adato, Michelle. ***Safety Second: The NRC and America's Nuclear Power Plants.*** Bloomington: Indiana University Press, 1987. 194 pages. ISBN 0253350344.

This is a critique of the Nuclear Regulatory Commission by the Union of Concerned Scientists. Contributors include James MacKenzie, Robert Pollard, and Ellyn Weiss. Generally, the writers maintain that the NRC has dodged many of the most serious safety issues, suffers from lax enforcement, is too closely tied to the industry it regulates, and does not encourage meaningful input from the public.

Alexanderson, E., and H. Wagner, eds. ***Fermi-I: New Age for Nuclear Power.*** La Grange Park, Ill.: American Nuclear Society, 1979. 454 pages.

This collection of papers is a window into the early history of nuclear power in the "Atoms for Peace" era, focusing on the proposal, design, development, operation, and eventual decommissioning of the Fermi-I experimental nuclear power plant. The authors also discuss management and organizational factors that contributed to the success of the project.

Allardice, Corbin, and Edward R. Trapnell. ***The Atomic Energy Commission.*** New York: Praeger, 1974. 236 pages. ISBN 0275554601.

This account of the history of the Atomic Energy Commission begins with Albert Einstein's letter of August 2, 1939, which first alerted the U.S. government to the potential of nuclear energy and weapons, and President Roosevelt's response in undertaking the Manhattan Project, with nuclear energy research under strict military secrecy. The birth of the AEC came with the Atomic Energy Act of 1946, which established a civilian nuclear power effort. The organization of the AEC and its relationship to its various constituencies (such as universities, researchers, and industries) is explored.

Balogh, Brian. ***Chain Reaction: Expert Debate and Public Participation in American Commercial Nuclear Power, 1945–1975.*** Cambridge, Mass.: Cambridge University Press, 1991. 340 pages. ISBN 0521372968.

This survey of the political history of nuclear energy in the United States looks at the emergence of an elite of experts who translated their authority into political power. The author argues that the decline of the power of the nuclear elite beginning in the 1970s was due to its failure to live up to its extravagant promises, not the ascendancy of anti-intellectualism in American culture. Activists with more coherent agendas gained better access to the American public, which could not relate to the arcane arguments of experts.

Birkland, Thomas A. ***After Disaster: Agenda Setting, Public Policy, and Focusing Events.*** Washington, D.C.: Georgetown University Press, 1997. 178 pages. ISBN 0878406530.

This is an academic study of how public awareness is shaped by disasters and how this dynamic in turn affects policymaking. The book begins with two chapters of theoretical background on the role of "focusing events" on agenda setting, and then devotes successive chapters to examples involving natural disasters, oil spills, and nuclear power plant accidents.

Campbell, John L. ***Collapse of an Industry: Nuclear Power and the Contradictions of U.S. Policy.*** Ithaca, N.Y.: Cornell University Press, 1988. 231 pages. ISBN 0801495008.

The author considers the U.S. nuclear industry as a case study in the interaction of markets and state intervention. The author argues (among other things) that the drive by companies for short-term profitability inevitably conflicts with the longer-term, comprehensive agenda sought by regulators, and that regulators in turn often fail due to poor planning and lack of coordination. The role of the antinuclear movement is also discussed, and comparisons are drawn with developments in Western Europe.

Cantelon, Philip L., Richard G. Hewlitt, and Robert C. Williams, eds. ***The American Atom: A Documentary History of Nuclear Policies from the Discovery of Fission to the Present.*** 2d ed. Philadelphia: University of Pennsylvania Press, 1991. 369 pages. ISBN 0812213548.

This is an updated edition of a collection of primary-source documents involving nuclear developments from the 1940s to the 1990s. Although the emphasis is on nuclear weapons development and proliferation issues, the new edition adds a chapter about the Chernobyl disaster.

Center for Strategic and International Studies. ***The Regulatory Process for NuFclear Power Reactors: A Review.*** Washington, D.C.: CSIS, 1999. 88 pages. ISBN 0892063564.

This report by a Center for Strategic and International Studies (CSIS) panel examines the operation of the Nuclear Regulatory Commission (NRC) and its regulation of nuclear plants. It examines thirteen issues arising out of the interaction between the NRC, the nuclear industry, and the public. The working group included representatives from the public sector, industry, and public interest groups. The report attempts to identify ways in which the nuclear industry can adapt to the increasingly competitive electrical power market.

Clarfield, Gerald H., and William M. Wiecek. ***Nuclear America: Military and Civilian Nuclear Power in the United States, 1940–1980.*** New York: Harper and Row, 1984. 518 pages. ISBN 0060153369.

This is a history of U.S. nuclear policy from the beginnings of government interest in nuclear research in 1940 to 1980, which saw the nuclear industry in decline following the Three Mile Island accident. Both nuclear weapons and civilian nuclear power developments are covered.

Cohn, Steven Mark. ***Too Cheap to Meter: An Economic and Philosophical Analysis of the Nuclear Dream.*** Binghamton, Albany: State University of New York Press, 1997.

A comprehensive analysis of the rhetoric, policy-making process, and economics of the struggle over the development of nuclear power. Includes an analysis of economic and energy supply forecasts, the effects of government subsidies, and environmental justifications based on global warming.

Duffy, Robert J. ***Nuclear Politics in America: A History and Theory of Government Regulation.*** Lawrence: University Press of Kansas, 1997. 304 pages. ISBN 0700608532.

This is a historical study of fifty years of nuclear regulation in the United States, focusing on the relationship between changing political trends and the development and revision of regulatory policies. The demise of the Atomic Energy Commission and its replacement by the Nuclear Regulatory Commission in the 1970s are explored, along with the effects of the widespread trend toward deregulation in the Reagan era.

Eckstein, Rick. ***Nuclear Power and Social Power.*** Philadelphia: Temple University Press, 1997. 191 pages. ISBN 1566394864.

The author describes and analyzes the U.S. nuclear power industry in terms of the institutions and individuals who exercise social power. Begins with an introduction to social power theory and a historical overview of nuclear power issues. Discusses the role of government regulation, corporate power, local politics, and the exercise of social power in a democracy. Controversies over the building of the Shoreham (New York) and Seabrook (New Hampshire) nuclear power plants are used as case studies. The author concludes that the ability of the nuclear industry to develop without regard for environmental or other consequences is typical of how power works in a society where corporations and ordinary citizens have vastly unequal power.

Gottfried, Kate-Louise D., and Gary Penn, eds. ***Radiation in Medicine: A Need for Regulatory Reform.*** Washington, D.C.: National Academy Press, 1996. 328 pages. ISBN 0309053862.

The authors, part of the Committee for Review and Evaluation of the Medical Use Program of the Nuclear Regulatory Commission, Institute of Medicine, evaluate the use of reactor-generated nuclear materials in medicine. This issue was raised in the 1990s by an accident that resulted in the death of a patient in Pennsylvania through an inadvertent high-dose radiation exposure. The book describes the uses of radiation sources and materials in medicine, the potential risks, the roles of regulatory agencies (particularly the NRC), and critiques of regulation from the health field and elsewhere. Recommendations for regulatory reforms are given.

Gwin, Louis. ***Speak No Evil: The Promotional Heritage of Nuclear Risk Communication.*** New York: Praeger, 1990. 200 pages. ISBN 0275934454.

The author, an experienced public relations executive and communications professor at Virginia Polytechnic, describes the history and gives critiques of the U.S. nuclear power industry's public relations campaigns from the mid-1950s through the 1980s. He suggests that "feel-good" PR campaigns that began with "Our Friend, the Atom" in the 1950s omit important information about dealing with possible emergencies and thus have increased the risks faced by neighbors of nuclear power plants. Government regulators pay little attention to the extent or quality of plants' public safety communication efforts. He suggests policy changes that would make risk communication more effective.

Hewlett, Richard G., and Oscar Anderson, Jr. ***The New World, 1939–1946*** (vol. 1 of *The History of the Atomic Energy Commission*). University Park: Pennsylvania State University Press, 1962. 766 pages. ISBN 087079471X.

This is the beginning of a detailed history of U.S. nuclear policy, with this first volume culminating in the Atomic Energy Act and the formation of the Atomic Energy Commission.

Hewlett, Richard G., and Francis Duncan. ***Atomic Shield, 1947–1952*** (vol. 2 of *The History of the Atomic Energy Commission).* Berkeley: University of California Press, 1990. 718 pages. ISBN 0520071875.

The authors' history of the AEC continues with the early organization and implementation of the agency as the United States took its first steps toward establishing a commercial nuclear power industry.

Hewlett, Richard G., and Jack M. Holl. ***Atoms for Peace and War, 1953–1961: Eisenhower and the Atomic Energy Commission*** (vol.

3 of *The History of the Atomic Energy Commission*). Berkeley: University of California Press, 1989. 696 pages. ISBN 0520060180.

The authors continue the history of the AEC as early nuclear power plants go into operation and the "Atoms for Peace" program becomes the centerpiece of an attempt to portray nuclear energy in a positive light.

Lilienthal, David E. ***The Atomic Energy Years: 1945–1950*** (vol. 2 of the author's *The Journals of David E. Lilienthal)***.*** New York: Harper and Row, 1964. 666 pages.

The author's second volume, based on diary entries from his years as AEC chairman, gives an insider's view of the formative years of the agency and a critique of decisions made at the time.

Makhijani, Arjun, and Scott Saleska, eds. ***The Nuclear Power Deception: U.S. Nuclear Mythology from Electricity "Too Cheap to Meter" to "Inherently Safe" Reactors.*** New York: Apex Press, 1999. 270 pages. ISBN 0945257929.

This extensive analysis parallels the development of reactor technologies with the development of public relations assertions about the safety of nuclear power. It begins with an introduction to reactor technology and the nuclear fuel cycle, then describes each type of reactor developed since the 1950s and the arguments made by proponents and critics concerning its safety. It next critiques the nuclear industry's attempt to recover public confidence by promoting "inherently safe" reactor designs and offers renewable energy and conservation as superior alternatives for our energy future. Includes an extensive appendix on nuclear physics and technology concepts.

A Shock to the System: Restructuring America's Electricity Industry. Washington, D.C.: Resources for the Future, 1996. 138 pages. ISBN 0915707802.

The future of nuclear power, like that of all forms of energy, is being impacted by the ongoing deregulation process that is bringing much greater competition to the electric power industry. The contributors to this volume discuss alternative implementations of deregulation and how they affect utilities, customers, and policymakers.

Walter, J Samuel. ***Containing the Atom: Nuclear Regulation in a Changing Environment, 1963–1971.*** Berkeley: University of California Press, 1992. 533 pages. ISBN 0520079132.

This survey describes the collision between a confident, rapidly expanding U.S. nuclear power industry and a growing public awareness and concern about environmental issues. The Atomic Energy Commission faced this challenge while confronting its own internal contradiction: The agency was charged both with promoting the widespread use of nuclear power and with regulating its safety. The author, a historian of the Nuclear Regulatory Commission, arrays copious details to show how regulation evolved during this tumultuous period.

Wellock, Thomas Raymond. ***Critical Masses: Opposition to Nuclear Power in California, 1958–1978.*** Madison, Wis.: University of Wisconsin Press, 1998. 333 pages. ISBN 0299158543.

This detailed account tells the stories of the individuals and groups that clashed over the proposed siting of nuclear plants in various parts of California during the 1960s and 1970s. The author is well suited to explore this topic because he has a background in nuclear engineering and a doctorate in history. His account explores the social values that propelled the antinuclear effort. The landscape for the conflict ranges from the tiny town of Wasco in the Central Valley to the huge, growing suburbs of Los Angeles that had a voracious demand for new sources of electrical power.

Welsome, Eileen. ***The Plutonium Files: America's Secret Medical Experiments in the Cold War.*** New York: Dial Press, 1999. 580 pages. ISBN 0385314027.

When scientists were working on the atomic bomb, they didn't know how hazardous working with plutonium would be. As revealed in this shocking exposé, scientists decided to find out by conducting medical experiments in which patients, without being informed, were injected with small quantities of plutonium. In other experiments, both soldiers and civilians were exposed to high levels of radiation, either not being informed or being offered grossly misleading explanations. Expanded from the author's Pulitzer Prize–winning newspaper series.

Winkler, Allan M. ***Life under a Cloud: American Anxiety about the Atom.*** Urbana: University of Illinois Press, 1999 (reprint of 1993 Oxford University Press edition). 290 pages. ISBN 0252067738.

This is a wide-ranging history and analysis of U.S. public opinion on nuclear issues. A three-part dynamic is described in which government officials, scientists, and cultural opinion-shapers (such as journalists) interact. Most often the government officials get their way unless the scientists or cultural community is aroused and mobilizes public opinion in opposition. The author concludes that citizen activists have enjoyed modest but important success in changing nuclear policies.

Case Studies

Aron, Joan B. ***Licensed to Kill: The Nuclear Regulatory Commission and the Shoreham Power Plant.*** Pittsburgh: University of Pittsburgh Press, 1997. 184 pages. ISBN 0822956497.

The author describes errors and misjudgments in the design, planning, and construction for the Shoreham nuclear power plant in Long Island, New York. The clashes between government regulators, the Long Island Lighting Company, and advocacy groups are recounted. Ultimately, the plant was abandoned without ever going on-line. Out of consideration of these events emerge fundamental issues of responsibility, oversight, and decisionmaking and the roles of federal, state, and local governments.

Beaver, William. ***Nuclear Power Goes On-line: A History of Shippingport.*** Westport, Conn.: Greenwood Press, 1990. 208 pages. ISBN 0313272441.

This book discusses the history of the Shippingport, Pennsylvania, nuclear power station, which went into operation May 26, 1958, ushering in the era of commercial nuclear power. The author believes that Shippingport was not only a pioneer but a prototype. In three decades of operation, the plant faced all the problems and issues that have become general concerns within the industry and among the public. Each stage of the plant's history, from design and concept to construction, training, and operation, is explored in detail. The relationship of Shippingport to the development of AEC reactor policy and the widespread adoption of light-water reactor technology are also discussed.

Bedford, Henry F. ***Seabrook Station: Citizen Politics and Nuclear Power.*** Amherst: University of Massachusetts Press, 1990. 224 pages. ISBN 087023711X.

The author recounts the bitter struggle that erupted in 1972 when the Public Service Company of New Hampshire decided to build twin nuclear power plants in a salt marsh near Seabrook. Although the company eventually won operating licenses for the plants, the struggle left lasting scars in the community. The narrative is very detailed and draws heavily on documents to explain the strategies and tactics of both sides in events stretching from 1972 to 1989. What emerges are a faulty regulatory system, dubious behavior by both proponents and opponents of the plant, and the need to change both regulation and corporate policy.

McCallioin, Kenneth F. ***Shoreham and the Rise and Fall of the Nuclear Power Industry.*** Westport, Conn.: Praeger, 1995. 256 pages. ISBN 0275942996.

This is a history of the development, crises, and decline of the U.S. nuclear power industry. The main case study used is that of the Shoreham nuclear power plant on Long Island. After emergency diesel generators broke down, plant officials covered up the safety hazards, whistle-blowers stepped forward, and the case eventually wound up as a civil RICO (racketeering) trial. The case is seen as emblematic of the mixture of incompetence, cover-up, and fraud that has turned the public against nuclear power and led to the abandonment of a number of partly finished or inoperable nuclear power plants.

Future Prospects

International Nuclear Societies Council (INSC). ***A Vision for the Second Fifty Years of Nuclear Energy.*** American Nuclear Society, 1996. 80 pages.

This report represents the consensus opinion of the INSC's worldwide membership on strategies that should be pursued for the worldwide development of nuclear power to meet the energy needs of the twenty-first century.

International Nuclear Societies Council. ***Worldwide Integrated View on Main Nuclear Energy Issues.*** International Nuclear Societies Council, 1999 (7 papers).

This is a collection of papers by working groups of the INSC. Titles: 1, Toward a Worldwide Consensus about Safety of Nuclear Reactors; 2, Achieving Public Understanding and Acceptance of

Nuclear Power; 3, Important Issues in Global Utilization of Nuclear Power; 4, Radioactive Waste; 5, Nuclear Nonproliferation; 6, Role of Risk Methods in the Regulation of Nuclear Power; 7, Low Doses of Ionising Radiation Incurred at Low Dose Rates. The papers are also available at the INSC website: http://www2s.biglobe.ne.jp/~INSC.

Morone, Joseph G., and Edward J. Woodhouse. ***The Demise of Nuclear Energy?: Lessons for Democratic Control of Technology.*** New Haven, Conn.: Yale University Press, 1989. 172 pages. ISBN 0300044496.

This book does not take sides or try to resolve the debate over nuclear power in the United States. Rather, it is a study of the relationship between emerging technologies, democratic institutions, and the political process. The different ways in which nuclear energy has been presented and challenged are surveyed. The authors conclude by suggesting reforms that might help our society better cope with the challenges of such technologies.

National Research Council Committee on Nuclear Power Development. ***Nuclear Power: Technical and Institutional Options for the Future.*** Washington, D.C.: National Academy Press, 1992. 215 pages. ISBN 0309043956.

This report deals with the virtual halt in growth of the U.S. nuclear industry, detailing the issues that must be addressed by both government and industry if a viable nuclear option is to be preserved for the future. The focus is on emerging new reactor technologies and how the alternatives compare with regard to safety, cost, and suitability. The effectiveness of current federal research programs is also evaluated.

Probst, Katherine, and Michael McGovern. ***Long-Term Stewardship and the Nuclear Weapons Complex: The Challenge Ahead.*** Washington, D.C.: Resources for the Future, 1998. 67 pages. ISBN 0915707977.

As a growing number of U.S. and former Soviet nuclear weapons are dismantled, policymakers are challenged to come up with a long-term program for the management of the resulting radioactive materials. This policy study outlines the parameters for a "long-term stewardship program" involving the cleanup of environmental contamination, environmental restoration, and safe monitored waste storage. The study calls for high-level coordination in the Department of Energy and the fostering of a public dialogue involving all "stakeholders," including the DOE, EPA, state agencies, local governments, tribal nations, citizens' groups, and private industry.

Ramsey, Charle B., and Mohammad Modarres. ***Commercial Nuclear Power: Assuring Safety for the Future.*** New York: John Wiley, 1998. 508 pages. ISBN 0471291862.

This is a comprehensive textbook covering all aspects of nuclear power plant operation in relation to accidents and safety practices. Topics include an overview of nuclear plants and the nuclear fuel cycle, types of health and environmental effects, control and safety systems, problems faced in accident prevention, management, regulatory oversight, and accident investigation. Written for a nontechnical audience, though some scientific background would be helpful.

Weinberg, Alvin. ***Continuing the Nuclear Dialogue: Selected Essays.*** La Grange, Ill.: American Nuclear Society, 1985. 204 pages. ISBN 0894485520.

A wide-ranging collection of almost forty years of reflections by a member of the group that developed the first nuclear chain reaction in Chicago in 1942. The author played a key role in reactor design for submarine propulsion, use of radioisotopes in medicine, and other applications of nuclear energy. He addresses issues raised by the Three Mile Island accident and public concern about nuclear power, and challenges nuclear critics to put their arguments in perspective.

International Issues

Development of Nuclear Power in Other Nations

Chu, David S. L. ***Power Development in a Developing Country: Experiences with Taipower.*** American Nuclear Society, 1989. 368 pages.

The former president of the Taiwan Power Company describes how his country built a modern electrical power system, largely based on nuclear power. He discusses the impact of this development on the environment and culture of Taiwan and argues that nuclear power has been an important factor in the nation's economic success in the decades following World War II.

Hecht, Gabrielle. ***The Radiance of France: Nuclear Power and National Identity after World War II.*** Cambridge, Mass.: MIT Press, 1998. 453 pages. ISBN 0262082667.

This extensive historical study places the development of the French nuclear power industry in the context of postwar French history. It suggests that a major impetus for the growth of nuclear power was France's desire to embrace such new technologies to forge a new identity in the postcolonial world—the seeking of a new "radiance," which is also the French word for "radiation." The author combines cultural and technical analysis to examine the reasons for choices of plant designs and the response of local communities to nuclear developments. She argues that planners and engineers consciously used their designs to serve a political and social agenda.

Jasper, James M. ***Nuclear Politics: Energy and the State in the United States, Sweden, and France.*** Princeton: Princeton University Press, 1990. 327 pages. ISBN 0691078416.

The author compares the development of nuclear energy policy in the United States, France, and Sweden since the oil crisis of the 1970s. The three countries ended up with quite different situations: In the United States, plant construction ground to an indefinite halt. Sweden built its planned plants but then decided to phase out nuclear energy by 2010. In France, however, nuclear plans were expanded even as the energy crisis eased. Jasper argues that economic and political factors are insufficient for understanding these different outcomes. Having interviewed one hundred policymakers, industry officials, and activists, he concludes that cultural factors and personal perspectives are key to understanding nuclear developments.

Marples, David R., ed. ***Nuclear Energy and Security in the Former Soviet Union.*** Boulder, Colo.: Westview Press, 1997. 177 pages. ISBN 0813390133.

This collection of papers by noted scholars discusses nuclear energy and weapons-related issues in the former Soviet Union. Topics include the Chernobyl cleanup effort and dealing with other accidents and waste dumps, safety problems in nuclear operations, nuclear smuggling, the transfer of nuclear weapons from Ukraine to Russia, and other foreign policy issues.

Nelkin, Dorothy. ***The Atom Besieged: Extraparliamentary Dissent in France and Germany.*** Cambridge, Mass.: MIT Press, 1981. 235 pages. ISBN 0262140349.

This is an analysis of grassroots antinuclear power movements in France and Germany during the 1970s. In France, particularly, the ruling interests embraced nuclear energy, leaving opponents to find alternative bases of political power.

Nordhaus, William D. ***The Swedish Nuclear Dilemma: Energy and the Environment.*** Washington, D.C.: Resources for the Future, 1997. 167 pages. ISBN 0915707845.

This study of the Swedish nuclear industry begins with historical background on the Swedish electrical industry and its role in the economy. It then looks at the debate leading up to a national nuclear referendum and the interparty agreement of 1991. The Swedish Energy and Environmental Policy (SEEP) Model is discussed, and the economic and safety considerations for Swedish nuclear power are analyzed. The book concludes with a discussion of alternative scenarios and consideration of the future of nuclear power in Sweden.

Robinson, Marilynne. ***Mother Country: Britain, the Welfare State, and Nuclear Pollution.*** New York: Farrar, Straus and Giroux, 1989. 261 pages. ISBN 0374213615.

The author charges that the British government–run Sellafield nuclear processing plant has for forty years routinely let deadly nuclear wastes leak into the Irish Sea and the surrounding environment. She recounts how her investigations were hampered by the Official Secrets Act. Her passionate account reaches back into five hundred years of British social and labor history to suggest that powerful interests have always been able to enjoy virtual impunity for such attacks on people's health and the environment.

Winnacker, K., and K. Wirtz. ***Nuclear Energy in Germany.*** La Grange Park, Ill.: American Nuclear Society, 1979. 356 pages. ISBN 0894480189.

This is a history of nuclear research and development in Germany. It begins with the wartime experiments under the Nazi regime, proceeds through the First Conference on the Peaceful Uses of Atomic Energy in 1955, and then looks at the implementation of various German nuclear programs. These developments are presented in a technical, political, social, and economic context.

Nuclear Proliferation

Allison, Graham T., Owen R. Coté, Jr., Richard Falkenrath, and Steven E. Miller. ***Avoiding Nuclear Anarchy: Containing the Threat of Loose Russian Nuclear Weapons and Fissile Material.*** Cambridge, Mass.: MIT Press, 1996. 295 pages. ISBN 026251088X.

The authors address the troubling vulnerability of the United States and other nations to terrorists who may be able to obtain or construct nuclear weapons more easily because of the chaos and lack of control in Russia's gigantic nuclear establishment. Although the main emphasis is on military weapons facilities, control of fissionable material is also a vital concern for civilian power plants. The book concludes with policy recommendations for meeting the challenge.

American Nuclear Society. ***Special Report on the Protection and Management of Plutonium.*** 1995, 94 pages. ISBN 089448561X.

This report surveys current issues arising from the large amount of plutonium that is being freed by the reduction of the U.S. and former Soviet nuclear arsenals, as well as plutonium recovered from spent fuel in commercial nuclear power plants. Includes executive summary and panel reports covering issues such as energy supply, economic factors, reactors and reprocessing of fuel, dismantling nuclear weapons, and regulatory requirements, as well as health, safety, environmental, and security concerns.

Baker, John C. ***Non-Proliferation Incentives for Russia and Ukraine.*** New York: Oxford University Press for the International Institute for Strategic Studies, 1997. 91 pages. ISBN 0198293712.

The author discusses the proliferation risks arising from the emergence of private nuclear industries (such as Minatom) in Russia and Ukraine. Focuses on the current incentives of the players and ways to create new incentives that would promote responsible behavior.

Beck, Peter. ***Prospects and Strategies for Nuclear Power: Global Boon or Dangerous Diversion?*** London: Earthscan Publications, 1994. 118 pages. ISBN 1853832170.

A researcher for the Energy and Environmental Programme, Royal Institute of International Affairs (London), examines the global issues arising from nuclear power and materials in the 1990s and beyond. He argues that it is important to pursue a viable nuclear option because of the likely need for a major power source that does not cause global warming. At the same time, current energy economics and the competitive environment do not make nuclear power an attractive investment, and problems with nuclear waste and proliferation have yet to be solved in a way satisfactory to the public. The author urges that new, safer technologies be pursued while there is still time.

Clark, William Jr., and Ryukichi Imai, eds. ***Next Steps in Arms Control and Non-Proliferation: Report of the U.S.-Japan Study Group on Arms Control and Non-Proliferation after the Cold War.*** Carnegie Endowment for International Peace and International House of Japan, 1996. 196 pages. ISBN 0870031058.

This report is from a conference of Japanese and American experts cosponsored by the Carnegie Endowment for International Peace and International House of Japan. Participants review thirteen issues related to nuclear nonproliferation in South and East Asia, including the possible proliferation threat arising from use of plutonium in commercial power reactors.

Goldschmidt, Bertrand. ***The Atomic Complex: A Worldwide Political History of Nuclear Energy.*** La Grange Park, Ill.: American Nuclear Society, 1982. 479 pages. ISBN 0894485504.

This review of international nuclear developments from the 1950s through the 1970s is now a good historical resource and background for decisions whose effects continue today. The role of industrial, governmental, and intergovernmental organizations is discussed in relationship to technical developments.

Harrison, Selig, ed. ***Japan's Nuclear Future: The Plutonium Debate and East Asian Security.*** Washington, D.C.: Carnegie Endowment for International Peace, 1996. 120 pages. ISBN 0870030655.

Having made a major commitment to nuclear power and the reprocessing of plutonium, Japan is faced with the need to prevent nuclear proliferation and become part of the solution rather than part of the problem. The book begins by examining the background of the early nuclear choices faced by Japan (especially regarding plutonium) and their ramifications. It then focuses on the implementation of the Japanese plutonium program. Both the economics and the risks (proliferation, diversion, and health impact) are examined in detail.

Jones, Rodney W. ***Tracking Nuclear Proliferation: A Guide in Maps and Charts, 1998.*** Washington, D.C.: Carnegie Endowment for International Peace, 1998. 327 pages. ISBN 0870031139.

This is the latest survey prepared by the Carnegie Endowment for International Peace, organized into maps of each country with detailed charts showing nuclear facilities (such as power plants and research reactors) that could be used to develop nuclear weapons, as well as nonproliferation measures currently in place. The book also examines emerging proliferation threats caused by such developments as political instability in post-Soviet Russia and policies of the Chinese government that may facilitate sales of nuclear materials to other countries.

Kahn, Julius N., ed. and comp. ***Atoms for Peace Manual: A Compilation of Official Materials on International Cooperation for Peaceful Uses of Atomic Energy, December 1953–July 1955.*** Senate Document No. 55, 84th Congress, first session. Washington, D.C.: U.S. Government Printing Office, 1955. 615 pages.

This is an important historical record containing documents from the early "Atoms for Peace" effort, representing an attempt to promote international use of atomic power.

Leeuwen, Marianne Van, ed. ***The Future of the International Nuclear Non-Proliferation Regime.*** Boston: M. Nijhoff, 1995. 326 pages. ISBN 0792334337.

This collection of papers evaluates the current status of nuclear nonproliferation with regard to the former Soviet Union, North

Korea, South Asia, and the Middle East. The available mechanisms for monitoring and control are analyzed in light of possible threats and risks in each area. Concludes with proposals for strengthening the nonproliferation regime.

Roberts, L. E. J. ***Nuclear Power and Public Responsibility.*** New York: Cambridge University Press, 1984. 143 pages. ISBN 0521247187.

The former director of the Atomic Energy Research Establishment in Harwell, England, offers a British perspective on nuclear issues.

Simpson, John, ed. ***Nuclear Non-Proliferation: A Reference Handbook.*** Detroit, Mich.: Gale Research, 1992. 406 pages. ISBN 0582096480.

This is a resource guide to nuclear proliferation issues, which involve the spread of the technology and materials that might enable countries or subnational groups that do not currently have nuclear weapons to gain such capability. Since fuel for reactors (and particularly plutonium or enriched uranium recycled from such fuel) is the key component to nuclear weapons, nonproliferation issues impinge directly on nuclear power.

Simpson, John, and Daryl Howard, eds. ***Future of the Non-Proliferation Treaty.*** New York: St. Martin's Press, 1995. 226 pages. ISBN 0333618572.

This collection of seventeen papers was presented at a conference in Southampton, England, April–May 1995. The conference participants reviewed the Treaty on the Non-Proliferation of Nuclear Weapons with regard to issues of safety, security, peaceful uses of nuclear power, and national and regional activities, making recommendations with regard to extending the treaty.

Sweet, William. ***The Nuclear Age: Atomic Energy, Proliferation, and the Arms Race.*** 2d ed. Washington, D.C.: Congressional Quarterly, 1988. 340 pages. ISBN 0871874660.

This is an overview and perspective on nuclear issues, including safety and proliferation, at the end of the 1980s, looking toward a possible "post nuclear age" and its legacy. (Readers should supplement this by reading articles or books about nuclear-related events of the 1990s, especially the fall of the Soviet Union and the need to dismantle Russian and American weapons stockpiles.)

Nuclear Disasters

Three Mile Island

Houts, Peter S. ***The Three Mile Island Crisis: Psychological, Social, and Economic Impacts on the Surrounding Population.*** University Park, Pa.: Pennsylvania State University Press, 1988. 118 pages. ISBN 0271006331.

This study examines the psychological, social, and economic impact of the Three Mile Island nuclear accident on the surrounding community.

Rees, Joseph V. ***Hostages of Each Other: The Transformation of Nuclear Safety since Three Mile Island.*** Chicago: University of Chicago Press, 1994. 238 pages. ISBN 0226706877.

This is a study of management and operations in U.S. nuclear plants and how they were affected by the near-disastrous Three Mile Island (TMI) accident in 1979. Based on more than a hundred interviews and detailed study of accounts and documents, the book concludes that TMI had a decisive impact in transforming shockingly shoddy safety practices into a surprisingly effective combination of self-regulation, a "culture of safety," and government oversight. A little-known organization, the Institute of Nuclear Power Operations (INPO), emerged to play an important role in improving nuclear safety practices.

Chernobyl

Flavin, Christopher. ***Reassessing Nuclear Power: The Fallout from Chernobyl.*** Washington, D.C.: Worldwatch Institute, 1987. 91 pages. ISBN 0916468763.

The author examines the profound effects of the Chernobyl disaster on public opinion about nuclear power and suggests that the public has turned decisively away from this technology. Takes an antinuclear viewpoint.

Marples, David R. ***Chernobyl and Nuclear Power in the USSR.*** New York: St. Martin's Press, 1986. 228 pages. ISBN 0312004575.

This book, written shortly after the disaster, provides one of the earliest detailed accounts of its causes and effects. The author describes the economic and energy problems that led the Soviet Union to rely heavily on nuclear energy. The way the authorities responded during the days immediately following the disaster reveals serious questions about the ability of the nation to manage its nuclear infrastructure and to prevent future accidents.

Marples, David R. ***The Social Impact of the Chernobyl Disaster.*** Edmonton: University of Alberta Press, 1988. 313 pages. ISBN 0888641419.

The author examines in detail the events of the first two years following the accident, using both Soviet sources and in-depth interviews with Soviet scientists, scholars, officials, and media people.

Medvedev, Zhores A. ***The Legacy of Chernobyl.*** New York: Norton, 1990. 352 pages. ISBN 039302802X.

A prominent Soviet physicist and noted dissident provides a detailed examination of the many effects of the world's worst nuclear disaster. After discussing the technicalities of what went wrong, he explains the impact of the accident of agriculture, the environment in general, human health, and the economy. These effects include the placing of more than 600,000 people (including 250,000 children) on lifetime monitoring for effects arising from their exposure to radiation. Nearly three million acres of agricultural land has been lost for decades. Medvedev argues that understanding the causes and impact of the disaster also requires an understanding of the politics of the Soviet era.

Yaroshinska, Alla. ***Chernobyl: The Forbidden Truth.*** Michèle Kahn and Julia Sallabank, trans. Lincoln: University of Nebraska Press, 1995. 135 pages. ISBN 0803299109.

This account describes how the response of authorities to the Chernobyl nuclear disaster was tragically inadequate. The immediate medical needs of the victims were not met, and poor recordkeeping made good follow-up care and research into the effects of nuclear exposure very difficult. The author suggests that this inadequate response is not confined to Chernobyl but is characteristic of the worldwide nuclear industry. Includes a foreword by physicist-activist John Gofman with his "Nine Essential Rules of Inquiry" and recommendations for effective and appropriate response to such situations.

Nuclear Waste Issues

Environmental Effects

Barlett, Donald L., and James B. Steele. ***Forevermore: Nuclear Waste in America.*** New York: Norton, 1986. 352 pages. ISBN 0393303071.

This is a detailed but impassioned account of nuclear waste and "the radioactive poisoning of America" through the actions of greedy corporations and government agencies that looked the other way.

Becker, C. Dale, ed. ***Aquatic Bioenvironmental Studies: The Hanford Experience, 1944–1984.*** New York: Elsevier, 1990. 306 pages. ISBN 0444886532.

This is a collection of papers dealing with the development of the Hanford, Washington, nuclear facilities and their impact on the nearby aquatic environment. Topics include assessment of reactor operations, studies of radiation dispersion, monitoring of radioactive effluents, transport and behavior of radionuclides, studies of radiation exposure to specific organisms, and thermal effects on the ecosystem.

Committee for the Compilation of Materials on Damage Caused by the Atomic Bombs in Hiroshima and Nagasaki. ***Hiroshima and Nagasaki: The Physical, Medical, and Social Effects of the Atomic Bombings.*** Eisei Ishikawa and David L. Swain, trans. New York: Basic Books, 1981. 706 pages. ISBN 0465029876.

This is a definitive study of the effects and consequences of the atomic bombings of Japan during World War II. The extent of these radiation effects makes them a focus for studies that attempt to measure the short- and long-term effects of varying degrees of radiation exposure. They are thus relevant to the development of radiation standards used in the nuclear power industry.

Gerber, Michele Stenehjem. ***On the Home Front: The Cold War Legacy of the Hanford Nuclear Site.*** Lincoln: University of Nebraska Press, 1997. 334 pages. ISBN 0803270682.

The author describes the troubled legacy of environmental and health effects from the Hanford Nuclear Reservation, which produced the huge amounts of fissionable material needed for America's now obsolete Cold War arsenal.

Glasstone, Samuel, and Walter H. Jordan. ***Nuclear Power and Its Environmental Effects.*** La Grange Park, Ill.: American Nuclear Society, 1980. 395 pages. ISBN 0894480243.

This is a comprehensive introduction to safety and environmental issues relating to nuclear power. Begins with fundamental principles of operation, then looks at the reactor licensing process, reactor safety and standards, natural radiation in the environment, biological effects of radiation, radiation arising from nuclear power operations, waste management, transportation of radioactive materials, and thermal effects.

Gould, Jay M., and Benjamin A. Goldman with Kate Millpointer. ***Deadly Deceit: Low-Level Radiation, High-Level Cover-Up.*** New York: Four Walls Eight Windows, 1991. 222 pages. ISBN 0941423352.

This exposé extensively documents what the author considers to be the pervasive and deadly effects of radiation that are ignored by the nuclear industry. Contains many anecdotes and proposals for reform. Some critics and former nuclear workers have objected to what they see as an overreliance on anecdotes as opposed to hard, quantifiable evidence.

Hendee, William R., and F. M. Edwards, eds. ***Health Effects of Exposure to Low-Level Ionizing Radiation.*** 2d ed. Philadelphia: Institute of Physics Publishers, 1996. 568 pages. ISBN 0750303492.

Twelve contributors discuss radiation sources, health effects (particularly cancer), radiation protection, perception of radiation hazards by workers and the general public, and ways to educate people about radiation. The interplay of physics, medicine, biology, and statistics is explored as methodology is explained.

Morgan, Karl Ziegler, and Ken M. Peterson. ***The Angry Genie: One Man's Walk through the Nuclear Age.*** Norman: University of Oklahoma Press, 1999. 218 pages. ISBN 0806131225.

These are the memoirs of Karl Ziegler Morgan, physicist at the Manhattan Project and Oak Ridge National Laboratory, who became one of the key founders of health physics as a discipline. He recounts radiation accidents and near disasters that led him to become a staunch critic of practices in the nuclear industry and to testify in the cases of *Silkwood v. Kerr-McGee Corporation* and *Allen v. United States.*

Schull, William J. ***Effects of Atomic Radiation: A Half-Century of Studies from Hiroshima and Nagasaki.*** New York: Wiley-Liss, 1995. 397 pages. ISBN 0471125245.

This is a definitive survey and account of the study of radiation effects. The author, a geneticist, served with the Atomic Bomb Casualty Commission beginning in 1949, conducting extensive long-term studies of the effects of radiation exposure on survivors of the atom bombings in Japan. His research now forms much of the basis for the presently accepted radiation standards. Recounting of personal experiences lends immediacy to the account.

Waste Disposal and Management

Benford, Gregory. ***Deep Time: How Humanity Communicates across Millennia.*** New York: Avon Books, 1999. 225 pages. ISBN 0380975378.

Managing radioactive waste that will remain dangerous for thousands of years requires a way to communicate with one's distant descendants across a time span potentially greater than that from the first cities (civilizations) to today. The author looks at the parameters and implications for such long-term communication. He suggests that we have already made a long-term impact on the planet and explores ways to use nature for intentional communication.

Herzik, Eric B., and Alvin H. Mushkatel, eds. ***Problems and Prospects for Nuclear Waste Disposal Policy.*** Westport, Conn.: Greenwood Press, 1993. 176 pages. ISBN 031329058X.

This collection of papers recognizes the key role that safe disposal of nuclear wastes will play in the future viability of nuclear power. The wide-ranging contributions cover such topics as the relationship between science, policies, and politics; legal issues surrounding waste disposal; the controversy over Yucca Mountain and possible alternatives to a centralized national waste depository; and monitoring and transportation of wastes.

Hevly, Bruce William, and John M. Findlay, eds. ***The Atomic West.*** Seattle: University of Washington Press, 1998. 286 pages. ISBN 0295977493.

This historical survey looks at how cultural perceptions of the American West as barren and desolate encouraged atomic planners to site many nuclear facilities in the western states. Planners tended to believe that nuclear pollution would have little effect in the sparsely populated West.

National Research Council. ***Management and Disposition of Excess Weapons Plutonium: Reactor-Related Options.*** Washington, D.C.: National Academy Press, 1995. 436 pages. ISBN 0309051452.

This report is from a panel discussing possible uses of nuclear reactors for converting the hundreds of tons of weapons-grade plutonium and highly enriched uranium left from the dismantling of many nuclear weapons by the United States and Russia. Different types of reactors and procedures are discussed as well as the vitrification of plutonium. The possible economic, security, and environmental consequences of the different procedures are analyzed and compared.

National Research Council Committee on Decontamination and Decommissioning of Uranium Enrichment Facilities. ***Affordable Cleanup? Opportunities for Cost Reduction in the Decontamination and Decommissioning of the Nation's Uranium Enrichment Facilities.*** Washington, D.C.: National Academy Press, 1996. 324 pages. ISBN 0309054389.

Presents the results of a study commissioned by Congress in the 1992 Energy Policy Act. Examines the existing plans and estimated costs for decontaminating and decommissioning the nation's three existing uranium enrichment facilities (Oak Ridge, Tennessee; Paducah, Kentucky; and Portsmouth, Ohio). Identifies ways to reduce these costs and discusses options for disposing of the large quantity of depleted uranium hexafluoride currently stored at these sites.

National Research Council Committee on Separations Technology and Transmutation Systems. ***Nuclear Wastes: Technologies for Separations and Transmutation.*** Washington, D.C.: National Academy Press, 1996. 592 pages. ISBN 0309052262.

This report discusses alternatives to the current method of disposing of spent fuel from light water reactors, which involves direct disposal of the mixed radionuclides into storage sites. Alternatives include separation and transmutation technologies that can extract the more dangerous components (such as plutonium) and thus convert wastes that are long-term hazards into low-level wastes or wastes with a short half-life. Technologies that extract the highly fissionable portion of wastes greatly reduce the bulk of wastes that need to be disposed of but present risks of nuclear weapons proliferation, which are also discussed.

National Research Council Panel on Coupled Hydrologic/Tectonic/Hydrothermal Systems at Yucca Mountain. ***Ground Water at Yucca Mountain: How High Can It Rise?*** Washington, D.C.: National Academy Press, 1992. 242 pages. ISBN 030904748X.

After a Department of Energy staff scientist suggested that a major upwelling of groundwater could flood the Yucca Mountain national nuclear waste storage facility and spread radioactive wastes sometime in the future, the DOE requested this study, which looks at the risk of such a flooding beneath Yucca Mountain.

The Nuclear Waste Primer: A Handbook for Citizens. Rev. ed. Washington, D.C.: League of Women Voters Education Fund, 1993. 170 pages. ISBN 1558212264.

This publication by a respected voter education group provides a balanced introduction to nuclear waste issues. Topics covered include sources and forms of wastes, radiation dangers, policy and legislative developments, and current status of storage facilities.

Research Reactor Aluminum Spent Fuel: Treatment Options for Disposal. Washington, D.C.: National Academy Press, 1998. 240 pages. ISBN 0309060494.

This book presents the findings of a technical working group (under Milton Levenson, principal investigator; and Kevin D. Crowley, study director) that evaluated various methods for processing spent aluminum fuel in research reactors. Direct disposal, highly enriched uranium dilution, and advanced treatment technologies are examined. Requirements are established, and a comparative cost analysis is included. The report concludes with observations and recommendations.

Walker, Charles A., Leroy C. Gould, and Edward J. Woodhouse. ***Too Hot to Handle? Social and Policy Issues in the Management of Radioactive Wastes.*** New Haven, Conn.: Yale University Press, 1983. ISBN 0783753101.

This is a good introduction to issues relating to nuclear waste storage and disposal, including technical background and an emphasis on analysis of public opinion and perceptions. Although almost two decades have passed since the book was published, the fundamental technical and sociopolitical issues in this arena have changed very little.

Miscellaneous Issues

Fusion Power

International School of Fusion Reactor Technology, Ninth Meeting: 1989, Erice, Italy. ***Safety, Environmental Impact, and Economic Prospects of Nuclear Fusion.*** Edited by Bruno Brunelli and Heinz Knoepfel. New York: Plenum Press, 1990. 352 pages. ISBN 0306435241.

Although commercially viable nuclear fusion power has yet to be achieved, the experts at this conference discussed and formulated criteria that should be insisted on for any system to be acceptable. Proposed standards cover radiological safety within the facility, environmental impact, and operating standards.

Nuclear Power in Space

Atomic Power in Space: A History. Washington, D.C.: U.S. Dept. of Energy, 1987. 180 pages. (DOE/NE/32117-H1).

Power systems based on the heat generated by radioactive materials have been used in a variety of space probes launched by the United States and Soviet Union. This publication recounts the development of such systems, which have provoked strong criticism by antinuclear activists.

Grossman, Karl. ***The Wrong Stuff: The Space Program's Nuclear Threat to Our Planet.*** Monroe, Me.: Common Courage Press, 1997. 270 pages. ISBN 1567511244.

This advocacy book argues that the U.S. space program has used expensive, dangerous nuclear power for space missions despite the availability of less costly, safe solar power. Specific missions are investigated, including the Cassini mission, which carries 72.3 pounds of plutonium fuel. The dramatic narrative uncovers what the author claims to be lies and obfuscations used by space officials who are determined to pursue the nuclear option despite any risks.

Other

Bernard, John A., and Takashi Washio. ***Expert Systems Applications within the Nuclear Industry.*** La Grange Park, Ill.: American Nuclear Society, 1989. 249 pages. ISBN 0894480340.

This book is a useful review of how expert systems (rules-based artificial intelligence) software can be applied in the design, management, and operation of research and commercial reactors. Almost three hundred different systems are reviewed, grouped into sixty applications. Despite the considerable advances in computer hardware since the book was written, much of the software remains "state of the art."

Institute of Medicine Committee on Biomedical Isotopes. ***Isotopes for Medicine and the Life Sciences.*** S. James Adelstein and Frederick J. Manning, eds. Washington, D.C.: National Academy Press, 1995. 144 pages. ISBN 0309051908.

This overview begins with a summary of the essential uses of radioisotopes in medicine and science; often there is no feasible alternative to the use of radioisotopes for applications such as tracing. Cutbacks in funding for Department of Energy laboratories have reduced the availability of such isotopes and endangered medicine and scientific research. The report describes education and training needed to sustain production of needed isotopes and the building of partnerships between industry, government laboratories, and universities, and the elements of a national isotope policy.

Majumdar, M. Catherine, ed. ***Artificial Intelligence and other Innovative Computer Applications in the Nuclear Industry.*** New York: Plenum Press, 1988. 910 pages. ISBN 0306429020.

This is a collection of the American Nuclear Society Topical Meeting on Artificial Intelligence and Other Innovative Computer Applications, held August 31–September 2, 1987, in Snowbird, Utah, which discussed applications of artificial intelligence and robotics principles to nuclear power plant systems. Topics include diagnostic and alarm systems, plant control systems, and the design of more effective interfaces to allow human operators to identify and react to crucial information.

National Research Council Committee on Nuclear Engineering Education. ***U.S. Nuclear Engineering Education: Status and Prospects.*** Washington, D.C.: National Academy Press, 1990. 180 pages. ISBN 0309042801.

The diminishing number of undergraduates majoring in nuclear engineering may mean that there will not be enough nuclear engineers to meet the needs of the nuclear industry, potentially impacting power production and safety. The report suggests what should be included in the nuclear engineering undergraduate curriculum, how to integrate it with other disciplines, and how to encourage greater enrollment in the field.

Suid, Lawrence H. ***The Army's Nuclear Power Program.*** New York: Greenwood Press, 1990. 130 pages. ISBN 0313272263.

Although many people distinguish nuclear weapons development and the military from civilian development of nuclear power, the U.S. Army also designed and developed plans to produce nuclear power at remote military bases. This book traces the development of this program from its struggles with the competing interests of the civilian Atomic Energy Commission to the building of a prototype nuclear power plant at Fort Greely and the program's eventual decline. Includes bibliographical essay.

Periodicals

Many of the governmental, industrial, and activist organizations listed in Chapter 5 publish newsletters or other periodicals, either in paper form or via their websites.

Nuclear science and engineering are also sometimes covered in general science magazines such as *Scientific American* and *Science;* accidents or other developments are covered in general newsmagazines. See a general periodical index database such as Infotrac for recent publications.

The following is a selection of other periodicals that might be of interest to readers, especially those pursuing a more scholarly or technical inquiry.

Annals of Nuclear Energy. 18 issues/year. ISSN 03064549. Elsevier Science. http://www.elsevier.nl/.

A scholarly publication with text and articles summarized in English, French, and German.

Atomic Energy Insights. Monthly. http://ans.neep.wisc.edu/~ans/point_source/AEI/AEI_home.html.

Pronuclear. Seeks to provide information about the value of nuclear energy and its potential uses. For example, a recent issue suggests that burying valuable fissionable material from no-longer-needed nuclear weapons is illogical and wasteful, when such material can be reprocessed into fuel and used to create energy.

The Bulletin of the Atomic Scientists. Bimonthly. http://neog.com/atomic/bullatomsci@igc.apc.org.

This magazine was founded in 1945 by scientists who had worked on the atomic bomb. Its famous "Doomsday Clock" gave a graphic depiction of the danger of nuclear catastrophe during the Cold War. The bulletin is particularly valuable for coverage of nonproliferation issues.

Current Bibliography on Science and Technology: Nuclear Engineering. Monthly. ISSN 00113263. Japan Science and Technology Corporation.

Bibliography in English.

Journal of Nuclear Science and Technology. Monthly. ISSN 00223131.

Published by the Atomic Energy Society of Japan; text in English.

Nuclear Awareness News. 2/year. E-mail: nucaware@web.net.

Published by the Nuclear Awareness Project, which seeks to educate people about nuclear developments in Canada.

Nuclear Energy Info. Monthly. Published by U.S. Nuclear Society.

Trade publication.

Nuclear Engineering and Design. 24 issues/year. ISSN 00295493.

Scientific/technical publication. Published by Elsevier Science, USA.

Nuclear Engineering International. Monthly. ISSN 00295507.

Trade publication. Text in English, summaries in French and German. Published by Wilmington Publishing, Ltd. http://www.energy@wilmington.co.uk.

Nuclear Engineering Monthly. ISSN 04334035.

Trade publication. Published by Industrial Daily News, Ltd.

Nuclear Europe Worldscan. Bimonthly. ISSN 10165975. http://www.new@to.aey.ch.

Sponsored by European Nuclear Society; text in English.

Nuclear Index. Monthly. ISSN 02710706.

Bibliographical. Sponsored by National Energy Researchers; published by McGraw-Hill Companies.

Nuclear India. Monthly. ISSN 00295523.

Published by India's Department of Atomic Energy. Text in English.

Nuclear Law Bulletin. Semiannual, plus supplements. ISSN 0304341X. http://www.oecd.org. E-mail: sales@oecd.org.

Sponsored by the Organization for Economic Cooperation and Development.

Nuclear Monitor. Monthly. ISSN 08893411. http://www.nirs.orgnirsnet@igc.apc.org.

Published by the Nuclear Information and Resource Service.

Nuclear Plant Journal. Bimonthly. ISSN 08922055.

Journal for nuclear power professionals. Current and back issues available on-line at http://www.npj.goinfo.com/.

Nuclear Reactor Safety. Monthly. ISSN 07352492.

Published by the U.S. Department of Energy.

Nuclear Regulatory Reports. Weekly. ISSN 03607690.

Published by the U.S. Nuclear Regulatory Commission.

Nuclear Resister. Quarterly. ISSN 08839875. http://www.nonviolence.org/nukeresister. E-mail: nukeresister@igc.org.

Antinuclear. Published by the "National No-Nukes Prison Support Collective."

Nuclear Science and Engineering. Monthly. ISSN 0029-5639.

Academic and technical. Published by the American Nuclear Society.

Nuclear Technology. Monthly. ISSN 00295450.

Subtitled "Applications for nuclear science, nuclear engineering, and related arts." Published by the American Nuclear Society.

Nuclear Waste Bulletin. Irregular. ISSN 10227040. http://www.oecd.org. E-mail: sales@oecd.org.

Sponsored by the Organization for Economic Cooperation and Development.

Nuclear Waste News. Weekly. ISSN 02762897. http://www.bpinews.com. E-mail: bpinews@bpinews.com.

Covers "generation—packaging—transportation—processing—disposal" of nuclear waste. Published by Business Publishers, Inc.

Progress in Nuclear Energy. 8/year. ISSN 01491970. http://www.elsevier.nl/.

"International review journal covering all aspects of nuclear energy." Published by Elsevier Science.

Radwaste Magazine. Bimonthly.

Magazine featuring articles discussing solutions to a variety of problems involving the disposal and management of all types of nuclear waste as well as environmental restoration. Published by the American Nuclear Society.

World Nuclear Industry Handbook. Annual trade publication. http://www.energy@wilmington.co.uk.

Published by Wilmington Publishing, Ltd.

7

Nonprint Resources

World Wide Web pages, databases, and other services accessed via the Internet have become a vital resource for all researchers from high school students to academics and industry professionals. This chapter begins by presenting a categorized listing of important and useful websites in the area of nuclear power. The sites are broken down into the following categories:

- General resource, databases, and archives—sites that provide a wide range of information and links
- Historical and documentary
- Regulatory, legal, and governmental information
- Technical information about nuclear power
- Radiation and nuclear waste
- Antinuclear groups

I have tried to choose websites from organizations and groups that are likely to remain in existence for some time. However, it is the nature of the Web that much of its content is ephemeral. If the researcher doesn't find a given site under the address listed, try submitting the site name to a search engine to see if it can be accessed under a newer address. It is also a good idea to begin one's research at one or two of the general resource sites, which tend to have the most up-to-date links to other sites and information. Note, also, that most of the organizations listed in Chapter 5 have websites that include news, background information, and links to related sites. Those websites are not included in this listing.

The following list also includes some Usenet news groups relating to nuclear issues. These groups can be read and posted to from most Web browsers or through a site such as www.dejanews.com.

Another important source for current publications is of course the library catalog. Many libraries now make their catalogs remotely accessible via the Internet. Libraries also often allow their users to access periodical databases and indexes such as InfoTrac. There are also document services such as Dialog's Source One/Un-Cover© that can be used to both search for and obtain articles.

This chapter concludes with a selection of audiovisual material on nuclear power–related subjects. A large quantity of such material was created in the 1970s and early 1980s. Although much is still useful, it may be hard to obtain. The selection has been largely confined to materials published during the 1990s. In general, the Internet seems to be supplanting traditional audiovisual media as a means of visual education.

Websites

General Resources, Databases, and Archives

Atomic Archive
http://www.atomicarchive.com/main.shtml

A comprehensive, well-organized guide to the development of the atomic bomb and its physical and political consequences. Includes biographies, key documents, graphics, and videos. Although it does not focus on peaceful uses of nuclear power, there is much useful general background for students in that field.

Encyclopedia Britannica Online
http://www.britannica.com

A search of this on-line encyclopedia site using the keywords "nuclear power" highlights websites, magazines, and books on the topic.

Frequently Asked Questions about Nuclear Energy
http://www-formal.stanford.edu/jmc/progress/nuclear-faq.html

This page by John McCarthy, noted professor and artificial intelligence researcher at Stanford University, explains various facets of nuclear energy from a pronuclear viewpoint and argues that nuclear energy is safe and is likely to be the only truly sustainable source of energy for our general electrical needs.

Index to Databases (International Atomic Energy Agency)
http://www.iaea.or.at/databases/dbdir/fulllist.htm

On-line index to a variety of searchable technical databases on a variety of subjects relating to nuclear power plants, the nuclear fuel cycle, waste management, radiation sources, and so on.

Links to Nuclear Energy Sites
http://www.phoenix.net/~nuclear/energy.html

This site has a collection of links divided into businesses, government agencies, universities with nuclear energy programs, utilities with nuclear plants, and "other" sites, including alternative energy sites.

Microsoft Encarta 2000

This CD-based encyclopedia has a good (though brief) overview of nuclear energy and technology and provides links to related websites.

N-Base Nuclear Information Service
http://www.n-base.org.uk/

This is a 10,000-record database of references to articles on British nuclear issues including power, fuel reprocessing and plutonium projection, waste transportation and storage, and environmental and health effects. It has a generally antinuclear emphasis.

Nuclear Energy Related Resource Notebook (Department of Energy, Idaho Operations Office)
http://www.id.doe.gov/doeid/academic/notebook.htm

Provides links to publications, websites, and contacts related to nuclear power issues.

Nuclear Plant Information Books
Nuclear Regulatory Commission
http://www.nrc.gov/AEOD/pib/pib.html

Interactive map of the United States showing nuclear power plants. Click to retrieve information about each plant's design, system characteristics, and emergency response plan.

Nuclear Power: Subject Bibliography (U.S. Government Printing Office)
http://www.access.gpo.gov/su_docs/sale/sb-200.html

Lists government publications and reports relating to nuclear power for sale by the Government Printing Office. Most are technical or regulatory in nature. Items can be ordered on-line.

NucNet: The World's Nuclear News Agency
http://www.nucnet.aey.ch/nucnet/

This news service covers current worldwide nuclear industry developments. The service requires a paid subscription for full access but offers a free "sampler" at its website.

Historical and Documentary

Chernobyl Children's Project Home Page
http://www.adiccp.org/contents.html

Home page for a charitable project based in Ireland. Includes photographs, poems, and paintings by children, and excerpts from the book *Children of Chernobyl.*

Facts about Chernobyl Disaster
http://www.belarusguide.com/chernobyl1/chfacts.htm

This site is part of a "virtual guide to Belarus" and contains extensive links to many documents of interest to students of the Chernobyl disaster and its aftermath. Includes previously secret official Russian documents.

Fifty Years from Trinity
http://www.seattletimes.com/trinity/

Website for a series in the *Seattle Times* on the history and consequences of the atomic age. Includes stories and descriptions about the Trinity and Nevada test sites and the Hanford, Washington, nuclear facilities. Links to related articles on nuclear leftovers, radiation, and atomic legacy.

International Conference: One Decade after Chernobyl
http://www.iaea.or.at/worldatom/thisweek/preview/chernobyl/

Website with materials from a conference held in Vienna, Austria, April 8–12, 1996, that summarizes the consequences of history's worst nuclear accident. Includes summaries of technical symposia, press conferences, releases, and other documents.

Nuclear Reaction: Why Do Americans Fear Nuclear Power?
http://www.pbs.org/wgbh/pages/frontline/shows/reaction/

Website for an episode of the PBS *Frontline* television documentary series. Includes readings, interviews, answers to frequently asked questions, and links to other nuclear-related sites. Interviews include activists such as Lynda Braasch and Joyce Corrardi, psychiatrist Robert DuPont (on "nuclear phobias"), power company CEO Jim Howard, Ralph Nader, former secretary of energy Hazel O'Leary, and Nobel Prize–winning nuclear physicist Glenn Seaborg.

Regulatory, Legal, and Governmental Information

DOE Pulse
http://www.doe.gov/news/newsprr.htm

A monthly on-line newsletter describing developments in the national laboratories run by the DOE.

Fact Sheet: Nuclear Power Plant Emergency
http://www.fema.gov/home/fema/radiolo.htm

This fact sheet from the Federal Emergency Management Agency (FEMA) has guidelines for measures to be taken by the public in the event of an emergency at a nuclear power plant. Includes information about warnings, evacuations, and sheltering in place.

Federal Nuclear Regulations
http://www4.law.cornell.edu/cfr/10cfr.htm#start

Links to Code of Federal Regulations. Title 10 deals with energy, including chapter 1 (the Nuclear Regulatory Commission) and chapters 2, 3, and 10 (Department of Energy). See also Title 32 (National Defense), Title 40 (Protection of the Environment), and Title 49 (Transportation).

Laws and Regulations, United States: Radiation Control
http://www.rmis.com/db/lawradia.htm

This page from a website of Risk Management Internet Services has links to summaries of state laws dealing with radiation. (This site requires a subscription for full access.)

Radiation-Related Rules, Regulations, and Laws
http://www.physics.isu.edu/radinf/law.htm

This page from the Radiation Information Network site provides an overview of federal agencies and their regulatory codes, as well as state and foreign regulatory agencies.

A Short History of Nuclear Regulation, 1946–1999
http://www.nrc.gov/SECY/smj/shorthis.htm

This page from the Nuclear Regulatory Commission website provides a useful outline summary of the development of nuclear regulation in the United States, including the history of the Atomic Energy Commission, the NRC's predecessor.

U.S. Code Chapter 23: Development and Control of Atomic Energy
http://www4.law.cornell.edu/uscode/42/ch23.html

An outline of the chapter of the U.S. Code dealing with nuclear regulation. Headings can be clicked on to view the corresponding code sections.

U.S. Department of Energy, Energy Information Administration
http://www.eia.doe.gov/fuelnuclear.html

Gives up-to-date statistics on U.S. nuclear plants, uranium processing, and the uranium market.

U.S. Government Nuclear Agencies
http://www.physics.isu.edu/radinf/feds.htm#top

This site, part of the Radiation Information Network, provides a guide to the many federal agencies that play a role in regulating nuclear materials and activities.

U.S. Nuclear Regulatory Commission
http://www.nrc.gov/nrc.html

Site for the agency of the Department of Energy that regulates commercial nuclear power operations. The site includes news, information, and reference archives.

U.S. Nuclear Regulatory Commission Information Digest
http://www.nrc.gov/NRC/NUREGS/SR1350/V10/index.html
(1998 edition)

An annual report presented on the Web in interactive form (search the NRC site for later years as available). It describes the functions of the agency and provides a variety of facts and statistics about the nuclear industry that it regulates.

Technical Information about Nuclear Power

ABCs of Nuclear Science
http://user88.lbl.gov/NSD_docs/abc/home.html

A web page at the Lawrence Berkeley Laboratory designed to introduce students (and interested adults) to the basics of nuclear structure and radiation. Includes a variety of simple experiments.

Atomic Physics on the Internet
http://www.plasma-gate.weizmann.ac.il/API.html

This extensive site provides international links to a variety of laboratories, institutes, academic departments, archives, and other resources of interest to scientists, researchers, and students. Also includes news about conferences and job openings and information about technical databases.

The Canadian Nuclear FAQ
http://www.ncf.carleton.ca./~cz725/

This site by Dr. Jeremy Whitlock provides information about nuclear power in Canada, particularly the CANDU (Canada Deuterium Uranium) reactor design, in question-and-answer format. It has a generally pronuclear perspective.

Control the Nuclear Power Plant (Demonstration)
http://www.ida.liu.se/~her/npp/demo.html

This site by Henrik Eriksson features an interactive simulation of a nuclear power plant that users can operate over the Web. The user attempts to manipulate valves and controls to counter one of several randomly generated failure sequences.

Department of Energy Lab Fact Sheets
http://www.er.doe.gov/production/er-07/page2a.html

This site provides links to fact sheets describing the operations of the nation's government-run nuclear research labs.

Energy Files: The Virtual Library of Energy Science and Technology
http://www.osti.gov/EnergyFiles/

This is a project of the Department of Energy Office of Scientific and Technical Information. The site provides a huge number of links and resources to energy-related topics as well as innovative ways for browsers to explore "subject pathways" such as "fission and nuclear technologies."

Fusion Energy
http://www.fusioned.gat.com/

This educational website is provided by General Atomics. It provides explanations, educational resources, and activities for students.

INEEL Tours Home Page (Idaho National Engineering and Environmental Laboratory)
http://www.inel.gov/resources/tours/tours.html

This site has information about tours of nuclear-related facilities in Idaho, such as the Experimental Breeder Reactor-1 (EBR-1), the Radioactive Waste Management Complex (RWMC), the Idaho Nuclear Technology and Engineering Center (INTEC), and the Test Reactor Area (TRA). The site also offers "virtual" (on-line) tours of some facilities.

Nuclear Engineering (Yahoo)
http://dir.yahoo.com/Science/Engineering/
Nuclear_Engineering/

The Yahoo! page has links to organizations and other sites related to nuclear engineering and nuclear power.

Virtual Nuclear Tourist: Nuclear Power Plants around the World
http://www.cannon.net/~gonyeau/nuclear/index.htm

This site by Joseph Gonyeau lets Web visitors "tour" a variety of nuclear facilities while learning how nuclear reactors and power plants work. More than 300 pages are interwoven with links to further information.

Radiation and Nuclear Waste

Atomic Atlas (Public Citizen)
http://www.citizen.org/cmep/AtomicAtlas/atlas.htm

Interactive, map-based guide to nuclear waste transportation routes and facilities.

Chelyabinsk: The Most Contaminated Spot on the Planet
http://www.logtv.com/chelya/

This Web page is devoted to a former Soviet plutonium and tritium production complex at which accidents (including a 1957 explosion) and the wholesale dumping of radioactive wastes exposed workers and people in the surrounding area to high doses of radiation.

Chernobyl: Ten Years of Radiological and Health Impact
http://www.nea.fr/html/rp/chernobyl/chernobyl.html

An assessment by the OECD Nuclear Energy Agency (France) of the health impact of the Chernobyl nuclear disaster.

Radiation and Health Physics Page
http://www.umich.edu/~radinfo/

Created by students at the University of Michigan, this site brings together resources relating to health physics and radiation protection. Links include introduction, professional resources, organizations and societies, and educational resources.

Radiation Information Network (Idaho State University)
http://www.physics.isu.edu/radinf/index.html

This site has many links to background information, organizations, regulations, news, databases, and other resources related to radiation and radiation protection.

The Virtual Depository of Radwaste Information
http://www.ourworld.compuserve.com/homepages/geodev/Link.htm

Maintained by "Geosciences for Development and the Environment," this site has links to radiation waste–related operations and activities around the world. There are also links to governmental, industry, and environmental groups.

Waste Link
http://www.radwaste.org/index.html

This site provides more than five thousand links relating to nuclear waste issues, including transportation, processing, and storage. The main page is organized by topic category and then "drills down" to lists of links.

Yucca Mountain (Environmental Protection Agency)
http://www.epa.gov/radiation/yucca/

This site provides information and resources about the Yucca Mountain nuclear waste disposal site, including frequently asked questions, background documents, and transcripts of public hearings.

Antinuclear

Anti-Nuclear Websites and Related Resources
http://www.prop1.org/prop1/azantink.htm

This site, maintained by the antinuclear peace group Proposition One, has links to many antinuclear groups. Although the focus is on nuclear disarmament, many groups related to nuclear power and waste are also included.

Greens (St. Joe Valley, South Bend, Indiana)
http://www.users.michiana.org/greens/editorial/mobile.htm

This site, by a local Green Party group, focuses on publicity and actions against nuclear waste transport trains in Europe and the United States, which they believe recklessly endanger the population and the environment. The site includes detailed maps of nuclear waste transportation routes.

no.nukes Greenpeace Nuclear Campaign Website
http://www.greenpeace.org/~nuclear/

This website is for a project of Greenpeace International dedicated to stopping the production and utilization of plutonium (and of nuclear power and technology generally). The site offers explanations of the nuclear fuel cycle, nuclear "hot spots" around the world, nuclear accidents and regulatory failures, and alternative, renewable forms of energy.

The Nuclear Guardianship Library
http://www.nonukes.org/ngl.htm

This site provides documents intended to help people prevent the further destruction of the environment through radioactive materials. On-line documents include background materials, a "Safe Energy Handbook," philosophical and ethical perspectives, political issues, technical issues, and health and safety. Materials were originally developed by the Nuclear Guardianship Project.

Nuclear Liabilities
http://www.users.massed.net/~agnews/

This antinuclear site highlights risks and dangers that the author believes will claim many lives in the future. It includes summaries of nuclear accidents and risk assessments as well as news stories.

Toward a Plutonium-Free Future
http://www.nonukes.org/

Provides information and action guides for antinuclear activists. A project of the Japanese INOCHI movement against "nuclear colonialism."

News Groups

alt.energy.nuclear

Nuclear energy, general discussion.

alt.engineering.nuclear

Nuclear engineering and technology, general.

alt.sci.physics.plutonium

Science and handling of plutonium.

clari.tw.nuclear

Nuclear power and waste (moderated discussion).

gov.us.fed.nrc.announce

Nuclear Regulatory Commission announcements (moderated discussion).

gov.us.topic.energy.nuclear

Nuclear power and radioactive materials (moderated discussion).

sci.physics.fusion

Nuclear fusion theory.

Audiovisual Materials

The following is a selection of audiovisual materials (videotapes and films). Some have been produced for the educational market, and others have been produced by industry or activist groups or were first shown as TV documentaries. Addresses for sources of audiovisual materials appear in a separate section at the end of this chapter.

Background and Overviews

Acceptable Risks?
Date: Unknown
Media: videocassette (VHS or U-matic)
Length: 16 minutes
Price: $149.00 (rental $75.00)
Source: Films for the Humanities and Sciences

Although the subject of this report is the risks of nuclear power and nuclear waste, it is approached from the larger perspective of how people assess and react to risks. The difficulty of obtaining reliable information and the effect of fear tend to make informed decisionmaking particularly difficult in this field.

A Door into the New World: Nuclear Energy and American Society
Date: 1993
Media: videocassette (VHS)

Length: 26 minutes
Price: $20.00 ($10.00 for NEI members)
Source: Nuclear Energy Institute

Presents scenes showing the development of nuclear power in the United States, without narration or a pro or con bias.

Half Lives
Date: 1996
Media: videocassette (VHS)
Length: 56 minutes
Price: $99.95
Source: The Nuclear Waste Documentary Project

This documentary presents a history and overview of the nuclear age starting with the Manhattan Project. The role of the government in both military and peaceful development of nuclear energy is discussed, as are the consequences of the Three Mile Island and Chernobyl disasters.

Haves and Have-Nots
Date: 1989
Media: videocassette (VHS)
Length: 60 minutes
Price: $29.95
Source: Annenberg/CPB Project

This documentary explores the relationship between the development of nuclear power infrastructure in Third World countries and the desire of leaders in some such countries to build nuclear weapons. This discussion of nuclear proliferation is part of a series on "War and Peace in the Nuclear Age."

The Nuclear Age
Date: 1996
Media: videocassette (VHS)
Length: 48 minutes
Price: $89.95
Source: Films for the Humanities and Sciences

A documentary history of the nuclear age from the atom bomb to the controversial use of nuclear power and the Chernobyl disaster. Includes both archival film footage and interviews with people who played key roles in nuclear developments. (This is part of a series called "History of the 20th Century.")

Nuclear Power
Date: 1994 (updated 1998)
Media: videocassette (VHS)
Length: 40 minutes
Price: $126.00 (includes book)
Source: Hawkhill Associates

This documentary covers the development of nuclear energy, with vivid footage from Los Alamos, Three Mile Island, Oak Ridge, and other important nuclear sites. Part 1 covers the history of nuclear power. Part 2 gives scientific background on nuclear power and discusses the pros and cons of the technology. Intended audience level is high school to college.

Nuclear Power Plant Safety: What's the Problem?
Date: 1997
Media: videocassette (VHS)
Length: 60 minutes
Price: $129.00 (rental $75.00)
Source: Films for the Humanities and Sciences

Shows how a nuclear power plant works and explains the numerous safety devices and procedures. Asserts that the nuclear industry has achieved a high level of safety.

Nuclear Power Production
Date: 1989
Media: videocassette (VHS) or film (16 mm)
Length: 25 minutes
Price: $525.00
Source: Handel Film Corporation

Presents an overview of nuclear energy, including how it differs from conventional fuels, how power is produced from nuclear fission, and the dangers of radiation to the environment. Compares the type of reactor used at Chernobyl with safer designs in the United States and looks at possible future reactor designs.

Nuclear Reaction
Date: 1997
Media: videocassette (VHS)
Length: 60 minutes
Price: $69.95
Source: PBS Video

Presents a diverse range of interviews that explore attitudes about nuclear power in several countries. Includes interviews with scientists, government officials, and citizens. Pulitzer Prize–winning historian Richard Rhodes provides a context for understanding the technical facts, likely risks, and people's psychological responses.

Radiation
Date: 1998 (updated)
Media: videocassette (VHS)
Length: 36 minutes
Price: $126.00 (includes book)
Source: Hawkhill Associates

An introduction to radioactivity and its effects, for high school and college students or adults. Part 1 explains how radiation was discovered, including the experiments of Becquerel and the Curies. Part 2 discusses the sources and effects of radiation encountered by people today.

Seabrook: Do We Need It?
Date: Unknown
Media: videocassette (VHS) or film (16 mm)
Length: 29 minutes
Price: $55.00
Source: WGBH Educational Foundation; distributed by PBS Video

This documentary describes the four-year battle over whether to complete the Seabrook, New Hampshire, nuclear plant. This struggle is placed in the larger context of "power"—both electrical and political—and the decisionmaking process. Opponents of the plant attack it as unsafe and as a generator of deadly wastes, and they propose that people use alternative forms of energy.

Antinuclear Advocacy

Amory Lovins
Date: ca. early 1990s
Media: videocassette (VHS)
Length: 29 minutes
Price: $49.95
Source: EnviroVideo

An episode of the *Enviro Close-Up* TV series. Features an interview with alternative energy advocate Amory Lovins, head of the Rocky Mountain Institute. Lovins warns that developing nations should avoid the nuclear path, which inevitably leads to the ability to make nuclear weapons. Nations should turn to safe alternative forms of energy. High school through adult.

Critical Mass, Voices for a Nuclear-Free Future
Date: 1996
Media: videocassette (VHS)
Length: 29 minutes
Price: $49.00
Source: EnviroVideo

A discussion by environmentalist and antinuclear activists including Helen Caldicott and Michio Kaku on the tenth anniversary of the Chernobyl disaster. Participants warn of the long-term medical consequences of nuclear power and of the international effort to promote the nuclear industry. The International Atomic Energy Agency is also criticized.

The End of Nuclear Power
Date: 1987
Media: videocassette (VHS)
Length: 55 minutes
Price: $29.95
Source: Willow Mixed Media, Inc.

Prominent scientists such as George Wald and Jay Gould, as well as activists and reporters, discuss the environmental and health effects of nuclear power, the effect on long-term death rates at Three Mile Island and Chernobyl, and the need to replace nuclear power with safe energy sources.

Fighting for Safe Energy
Date: Unknown
Media: videocassette (VHS)
Length: 29 minutes
Price: $49.95
Source: EnviroVideo

An episode of the *Enviro Close-Up* TV series. Includes interviews with activists who are fighting against nuclear power and working to create energy alternatives. Subjects include Annette Larson of the New England Coalition on Nuclear Pollution, Guy Chichester of Greens USA, and Kurt Ehrenberg of the Energy America Education Fund. Suitable for high school, college, or adult audiences.

Judith Johnsrud and Dr. Donnell W. Boardman
Date: Unknown
Media: videocassette (VHS)
Length: 29 minutes
Price: $49.95
Source: EnviroVideo

Presents interviews with Judith Johnsrud, director of the Environmental Coalition on Nuclear Power, and Dr. Donnell W. Boardman, founder of the Center for Atomic Radiation Studies. They discuss the impact of radiation on health, including cancer.

Life, Death, and the Nuclear Establishment
Date: 1990s
Media: videocassette (VHS)
Length: 29 minutes
Price: $49.95
Source: EnviroVideo

Investigative reporter Karl Grossman gives an exposé of who is behind the push in the 1990s to revive nuclear power. He highlights the danger of nuclear power by pointing to government documents that estimate the number of people who would be killed in a core meltdown at each nuclear plant.

Michael Mariotte, Nuclear Information and Resource Service
Date: ca. early 1990s
Media: videocassette (VHS)
Length: 29 minutes
Price: $49.95
Source: EnviroVideo

In this episode of the *Enviro Close-Up* TV series, Michael Mariotte of the Nuclear Information and Resource Service is interviewed. He reviews the current state of nuclear power in the United States and identifies the interests and forces that are seeking to revive the declining industry.

Michio Kaku on Nuclear in Space
Date: 1994
Media: videocassette (VHS)
Length: 80 minutes
Price: $49.95
Source: EnviroVideo

Presents a lecture by Michio Kaku, professor of nuclear physics at the City University of New York. The lecture was presented at a conference organized by the Global Network Against Weapons and Nuclear Power in space. Kaku points out the dangers of deploying not only weapons but supposedly peaceful nuclear technologies in space. (EnviroVideo Lecture Series.)

Millstone at the Crossroads
Date: Unknown
Media: videocassette (VHS)
Length: 29 minutes
Price: $49.95
Source: EnviroVideo

Contains interviews with activists who are exposing unsafe conditions at three Millstone nuclear power plants in Connecticut. Includes Pete Reynolds (who was fired for exposing Millstone's dangers), Geri Winslow of the Citizens Regulatory Commission, and energy expert Jim Newberry.

No Nukes Is Good Nukes
Date: Unknown
Media: videocassette (VHS)
Length: 26 minutes
Price: $59.00
Source: Hawkhill Associates

An interview with "soft energy path" expert Amory Lovins. He argues that nuclear power is both unsafe and unnecessary. Existing plants should be dismantled in favor of promoting energy efficiency and alternative forms of energy.

The Nuclear Expansion in Asia and Australia
Date: 1998
Media: videocassette (VHS)
Length: 29 minutes

Price: $49.95
Source: EnviroVideo

Reports how a nuclear industrial complex is growing in Australasia, with an increase in uranium mining in Australia feeding a growing number of nuclear power plants in Asia. The report features interviews with M. V. Ramana of MIT's Defense and Arms Control Studies Program and Tilman A. Ruff of International Physicians for the Prevention of Nuclear War. This is an episode of the *Enviro Close-Up* TV series.

Nuclear Power: Dangerous Energy
Date: 1991
Media: videocassette (VHS)
Length: 17 minutes
Price: $19.95
Source: Greenpeace, USA, Public Information Department

Presents a hard-hitting critique of nuclear energy as dangerous and not a real solution to the problem of global warming. The presentation includes a number of international perspectives and a focus on the meaning of the Chernobyl disaster.

Prairie Island Coalition
Date: Unknown
Media: videocassette (VHS)
Length: 29 minutes
Price: $49.95
Source: EnviroVideo

Reporter Karl Grossman interviews George Crocker of the Prairie Island Coalition, an environmentalist group that opposes nuclear power. The interview focuses on the issue of whether a nuclear waste storage facility should be built in Prairie Island, Minnesota. This was an episode of the *Enviro Close-Up* TV series.

The Push to Revive Nuclear Power
Date: ca. 1992
Media: videocassette (VHS)
Length: 29 minutes
Price: $49.95
Source: EnviroVideo

This film looks at the movement to revive the U.S. nuclear industry, as expressed in the U.S. National Energy Strategy. It also looks at alternative forms of energy that could meet national needs without the risks inherent in nuclear power. A number of antinuclear and alternative energy experts are interviewed, including Robert Pollard of the Union of Concerned Scientists, Amory Lovins, and Michael Mariotte. The film won an Honorable Mention at the 1992 Environmental Film Festival.

The Transportation of Nuclear Materials
Date: Unknown
Media: videocassette (VHS)
Length: 40 minutes
Price: $149.00 (rental $75.00)
Source: Films for the Humanities & Sciences

This documentary is an exposé of the transportation of nuclear waste by sea, a little-known (and perhaps cover-up) activity that raises many questions about safety and just what is being done with these materials.

Pronuclear Advocacy

Nuclear Power: Energy for the Future
Date: Unknown
Media: videocassette (VHS)
Length: 33 minutes
Price: $59.00
Source: Hawkhill Associates

An interview with pronuclear physicist Bernard Cohen, who argues that nuclear power is the best choice for energy in the future, both economically and environmentally.

Time Out for Science: Benefits and Uses of Nuclear Energy
Date: 1990
Media: videocassette (VHS) and printed materials
Length: 24 minutes
Price: $95.00
Source: United Learning

This is a presentation by the American Nuclear Society that details many ways in which advances in nuclear technology have benefited society.

Chernobyl and Other Nuclear Accidents

Changes
Date: 1979
Media: videocassette (VHS or U-matic)
Length: 29 minutes
Price: unknown
Source: Penn State Media Sales

Presents an intimate visit with seven people who live and work near the Three Mile Island nuclear plant. They describe how the accident at the plant changed their lives in various ways.

Chelyabinsk: The Most Contaminated Spot on the Planet
Date: 1995
Media: videocassette
Length: 58 minutes
Price: $350.00 (rental $75.00)
Source: Filmaker's Library

Interviews with farmers and other residents of an area victimized by decades of spills and contamination from the Soviet nuclear weapons industry, with results that may dwarf even the aftermath of Chernobyl. The collapse of the Soviet Union lifted the veil of secrecy and is revealing the true extent of the "most contaminated spot on the planet." The film has won numerous prizes for environmental and documentary filmmaking.

Chernobyl: Chronicle of Difficult Weeks
Date: 1987
Media: videocassette (VHS)
Length: 54 minutes
Price: $59.95
Source: The Video Project

This Russian film by Vladimir Shevchenko (with English subtitles) was banned by the Soviet Union but eventually was aired at the Glasnost Film Festival. It reveals the excruciating attempts of workers to contain the Chernobyl disaster.

Incident at Brown's Ferry
Date: 1977
Media: videocassette (VHS or U-matic) or film (16 mm)
Length: 58 minutes

Price: $19.95
Source: Time Life Education/Turner Le@rning/Time Life Video

This episode from the PBS *Nova* series examines the 1975 fire at the Brown's Ferry, Alabama, nuclear power plant, which revealed shortcomings in worker training and safety procedures. The broader issue of the safety of the U.S. nuclear industry is also discussed.

Nuclear Cover-Up: Chernousenko on Chernobyl—We Shall Die in Silent Ways
Date: unknown
Media: videocassette (VHS)
Length: 30 minutes
Price: $29.95
Source: The Video Project

Presents a Russian scientist's exposé of what really happened at Chernobyl. He explains how the Soviet government quickly covered up the true extent of the disaster. He believes that nuclear power cannot be made safe and that it poses a deadly threat to the planet.

Sixty Minutes to Meltdown
Date: 1984
Media: videocassette (VHS or U-matic) or film (16 mm)
Length: 84 minutes
Price: $19.95
Source: Time Life Education/Turner Le@rning/Time Life Video

A "docudrama" portraying the crucial minutes at the Three Mile Island nuclear plant, where malfunctions and human errors began to pile up, leading to a partial meltdown of the reactor core. The consequences of the accident for the safety and economics of nuclear power are then examined.

Three Mile Island Revisited
Date: ca. 1993
Media: videocassette (VHS)
Length: 30 minutes
Price: $49.95
Source: EnviroVideo

This documentary calls misleading the nuclear power industry's claims that no one died as a result of the Three Mile Island accident. Cancer deaths and birth defects have increased among residents of the area, and the utility has been quietly paying damage claims.

Nuclear Waste and Environmental Effects

Deafsmith: A Nuclear Folktale
Date: 1990
Media: videocassette (VHS)
Length: 43 minutes
Price: $79.00 ($45.00 rental)
Source: The Video Project

This documentary reports on the struggle of residents of Deafsmith County, Texas, to stop a Department of Energy plan to bury nuclear wastes in their farmland. Contains interviews with both resident-activists and DOE officials.

Fitting the Pieces: Managing Nuclear Waste
Date: 1985
Media: videocassette (VHS)
Length: 25 minutes
Price: free loan
Source: Nuclear Energy Institute

This presentation introduces high-level nuclear waste and uses animated sequences to show how it is shipped and then stored deep underground in a geologic repository.

Nuclear Nightmare Next Door
Date: 1988
Media: videocassette (VHS or U-matic)
Length: 52 minutes
Price: $149.00 (rental $75.00)
Source: Films for the Humanities and Sciences

Originally broadcast as an episode of CBS *48 Hours,* this documentary looks at the battle over proposed nuclear waste sites and the citizens who have organized to oppose putting the waste in their neighborhood.

Radioactive Reservation
Date: 1996
Media: videocassette (VHS)

Length: 52 minutes
Price: $295.00 (rental $75.00)
Source: Filmaker's Library

The U.S. government is trying to set up a series of "Monitored Storage Retrieval" nuclear waste sites in remote areas—areas such as Native American tribal lands. The government offers what appears to some tribal leaders to be "easy money" in exchange for storing waste on their land. Native American reservations facing such choices include the Paiute Shoshone in Oregon and the Cosiute near Salt Lake City, Utah. The film also looks at Apaches, Navajos, and Pueblos in New Mexico and Nevada who have had to face the results of nuclear testing for more than fifty years. The core of the film is statements by Native Americans themselves, who protest what they see as just the latest chapter in the exploitation of their people for profit.

Uranium
Date: 1990
Media: videocassette (VHS) or film (16 mm)
Length: 48 minutes
Price: $49.00 (rental $25.00)
Source: Bullfrog Films, Inc.

This film, produced by the Film Board of Canada, is an examination of the long-term environmental effects and social impact of uranium mining, specifically as they affect the miners and native peoples of the United States and Canada who live in areas near mines. The mines also threaten the spiritual lives of Native Americans, to whom much of the land involved is sacred. The desirability of continuing to mine uranium is questioned.

What'll We Do with the Waste When We're Through?
Date: 1987
Media: videocassette (VHS) or film (35 mm)
Length: 12.5 minutes
Price: $37.50
Source: United Learning

A presentation by the American Nuclear Society that explains what high-level nuclear waste is and how it can be disposed of safely.

Sources for Audiovisual Materials

Annenberg/CPB Project
P.O. Box 2345
South Burlington, VT 05407-2345
Phone: (800) 532-7637, (802) 862-8881, (802) 864-9846
E-mail: info@learner.org
Website: http://www.learner.org

Associated Press
50 Rockefeller Plaza
New York, NY 10020

Bullfrog Films, Inc.
P.O. Box 149
Oley, PA 19547
Phone: (800) 543-3764, (610) 779-8226, (610) 370-1978
E-mail: bullfrog@igc.org
Website: http://www.bullfrogfilms.com

EnviroVideo
P.O. Box 311
Ft. Tilden, NY 11695
Phone: (800) 326-8846, (718) 318-8045
E-mail: envirovideo@earthlink.net
Website: http://home.earthlink.net/~envirovideo/

Filmaker's Library, Inc.
124 E. 40th St., Ste. 901
New York, NY 10016
Phone: (800) 555-9815, (212) 808-4980, (212) 808-4983
E-mail: info@filmakers.com
Website: http://www.filmakers.com

Films for the Humanities and Sciences
P.O. Box 2053
Princeton, NJ 08543-2053
Phone: (800) 257-5126, (609) 275-1400, (609) 275-3767
E-mail: custserv@films.com
Website: http://www.films.com

Greenpeace USA, Public Information Department
1436 U St. NW

Washington, DC 20009
Phone: (202) 319-2444

Handel Film Corporation
8787 Shoreham Ave., Ste. 609
Los Angeles, CA 90069
Phone: (800) 395-8990, (310) 652-3887

Hawkhill Associates, Inc.
125 East Gilman St., P.O. Box 1029
Madison, WI 53701-1029
Phone: (800) 422-4295, (608) 251-3934, (608) 251-3924
Website: http://www.hawkhill.com

Nuclear Energy Institute
Website: http://www.nei.org

The Nuclear Waste Documentary Project
8505 Carter Mill Rd.
Knoxville, TN 37924
Phone: (423) 933-8233

PBS Video/Public Broadcasting Service
1320 Braddock Pl.
Alexandria, VA 22314-1698
Phone: (800) 344-3337, (703) 739-5380, (703) 739-5269
E-mail: www@pbs.org
Website: http://shop2.pbs.org/pbsvideo.default.asp

Penn State Media Sales
118 Wagner Bldg.
University Park, PA 16802
Phone: (800) 770-2111, (814) 863-3102, (814) 865-3172
E-mail: mediasales@cde.psu.edu
Website: http://www.mediasales.psu.edu

Time Life Education/Turner Le@rning/Time Life Video
P.O. Box 85026
Richmond, VA 23285-5026
Phone: (800) 449-2010, (804) 261-1300, (800) 449-2011

backfill In burying nuclear waste, the backfill is the soil or other material that is filled in around the waste canisters after they are placed in the ground.

background dose The amount of radiation received by a person from natural sources. It varies with altitude and the presence of natural sources of radioactivity (such as radon).

background radiation Radiation that either comes from space (the sun or cosmic rays) or is emitted by substances that occur naturally in the environment (such as traces of radioactive materials in construction materials and radon gas).

base load The normal electrical demand that a power plant is designed to supply.

base loaded A power plant maintained at its maximum capacity because it is the most economical source of power in the area.

baseline information Social, demographic, and environmental information gathered before siting a nuclear or other energy facility.

becquerel A unit of radioactivity equal to one atomic disintegration per second. A very tiny quantity equal to about twenty-seven picocuries (billionths of a curie).

beta particle A negatively charged particle that can be emitted through radioactive decay; essentially, a freely moving electron. Beta radiation has short range and limited penetrating power, so it is generally dangerous only if emitted by a source close to or inside the body.

boiling water reactor (BWR) A nuclear reactor in which water boils as it passes through the core. The boiling water serves both as a coolant (by carrying heat away from the reactor core) and as a source of steam to run a turbine to generate power. The steam is radioactive, so it must be contained and shielded within the turbine.

breeder reactor A nuclear reactor in which the neutrons resulting from fission are used to create more new fissionable atoms than were used in the reactor. The reactor thus "breeds" its own fuel.

British thermal unit (BTU) A unit of heat energy; the amount of heat required to raise the temperature of one pound of water by one degree Fahrenheit.

urn-up The amount of energy produced in a power reactor per unit of iginal nuclear fuel. Usually expressed as megawatt-days per metric

er dose The dose of radiation that if spread through a population pected to cause an additional cancer death. The "official" estimate of ncer dose is about 2,000 rem, but some critics believe it may be as 120–150 rem.

United Learning
6633 W. Howard St.
Niles, IL 60648-9990
Phone: (800) 424-0362

The Video Project
200 Estates Dr.
Ben Lomond, CA 95005
Phone: (800) 475-2638, (831) 336-0160, (831) 336-2168
E-mail: videoproject@videoproject.org
Website: http://www.videoproject.org

WGBH/WGBH Educational Foundation
125 Western Ave.
Boston, MA 02134
Phone: (617) 492-2777, (617) 787-0714
E-mail: feedback@wgbh.org
Website: http://www.boston.com/wgbh

Willow Mixed Media, Inc.
P.O. Box 194, Lennox Ave.
Glenford, NY 12433
Phone: (914) 657-2914

Glossary

This glossary briefly defines scientific and technical terms involved with nuclear physics, nuclear power technology, and regulation.

absorbed dose The amount of energy deposited in a given weight of biological tissue. The units of measure are the rad and the gray.

activity The rate at which a sample of radioactive material decays, measured in curies or becquerels.

adsorption The transfer of dissolved materials (such as radionuclide contamination in water) to a solid surface (such as a rock formation).

alpha particle A positively charged particle made up of two protons and two neutrons that can be emitted in the radioactive decay of som heavy atoms. Because it is relatively slow, an alpha particle cann penetrate clothing or even the outer layer of skin. It is dangerous if inhaled, ingested, or emitted by some source that has entere body.

aquifer Geological formations that can hold or carry water e as sand, gravel, or some kinds of rock. Nuclear waste stored an aquifer might get into the water supply.

atom The basic component of all chemical substanc made up of protons, neutrons, and electrons.

atomic energy *see* ***nuclear energy***

atomic mass The total number of protons and n the nucleus of an atom. For example, the most co has a total of 238 protons and neutrons and thu

atomic number The number of protons or determines what element an atom represen 92 protons and electrons has an atomic n ment uranium.

canister The outer metal container in which glassified high-level radioactive waste or spent fuel rods are placed.

cask A container used to transport and shield waste canisters or other radioactive materials during shipping.

chain reaction The process by which one nuclear fission emits neutrons that in turn cause other fissions, resulting in a self-sustaining release of nuclear energy.

chemical reaction A process in which one or more chemical substances interact, creating different substances and using or releasing energy. Unlike nuclear reactions, chemical reactions do not affect the cores of atoms and do not change one element into another.

cladding Material (such as steel or a metal alloy) that is used to encapsulate and shield pieces of radioactive fuel in a reactor.

cold fusion Hypothetical fusion reaction achieved by infiltrating atoms into a tightly packed crystal lattice rather than through the use of high temperature and pressure. Failure to replicate the experiments of the late 1980s has led most scientists to reject the existence of cold fusion.

colloid Small particles (10^{-9} to 10^{-6} meters in size) that are suspended in a solvent such as water. Natural colloids arise from clay minerals in contact with water, but radioactive waste can also escape into the environment in this form.

commissioning The process of making a nuclear power plant operational, which involves a lengthy testing and regulatory process.

condenser The pipes and other apparatus that allow steam to cool and turn back into water.

containment building The thick, steel-reinforced concrete building that surrounds the pressure vessel of a nuclear reactor. It represents the final barrier to the release of radioactive materials into the atmosphere.

containment system The layers of packaging and other systems designed to prevent radioactive materials from escaping into the environment during shipping.

contamination Radioactive material that has gotten into an undesired location, such as through leaks, leaching, or some other process.

control rods Long, thin rods made of a material that absorbs neutrons. Inserting rods into the reactor reduces the number of neutrons that can cause fission and thus brings the nuclear chain reaction to a halt.

control room The room in a nuclear power plant from which all operations are monitored and controlled through a large array of gauges, indicators, and controls.

coolant A fluid (usually water) that is circulated through the core of a nuclear reactor to remove the heat generated by the fission process. Some reactors use liquid metal (sodium) or a gas as a coolant.

cooling pond A pool or body of water used to dissipate heat from power plant operations through evaporation.

cooling system The system of pumps, pipes, and so on that carries coolant through a nuclear power plant.

cooling tower A tall, usually funnel-shaped structure that removes heat from water that has condensed from steam in the condenser. Cooling the water before putting it back in the environment prevents damage to wildlife.

core The central part of a nuclear reactor where the fission chain reaction takes place. It contains fuel rods, a moderator, and control rods. A coolant is continuously circulated through the core to remove heat, which is used to generate electric power.

cosmic rays High-energy particles from space that make up much of people's background radiation exposure.

critical The condition where a nuclear reaction is self-sustaining because the neutrons released by each fission can (on the average) trigger one new fission.

critical mass The minimum amount of fissionable material that for a given geometric arrangement allows for a self-sustaining chain reaction.

curie A unit of radioactive intensity equivalent to a gram of radium, representing thirty-seven billion atomic disintegrations per second.

decay The disintegration of an unstable atomic nucleus, resulting in the emission of charged particles and/or photons (gamma rays).

decommission Removal from service of a nuclear power plant that has ended its service life. Includes the dismantling and safe disposal of the reactor vessel and the reduction of radioactivity at the site so that it is available for unrestricted use.

Decon Nuclear Regulatory Commission term for decommissioning a nuclear plant by promptly decontaminating it and restoring the site to normal use.

decontamination The process of removing radioactive materials that have become mixed with the environment, a worker's clothing, or other materials.

defense in depth The design philosophy for nuclear facilities in which several different barriers are placed, each believed sufficient to prevent an escape of radioactive materials. For example, American reactors have both an inner containment and a strong outer building.

depleted uranium Uranium left over from the enrichment process that segregates and concentrates uranium-235. The depleted uranium is thus much lower in uranium-235 than natural unprocessed uranium. Because of its density, depleted uranium is used to make shells or bullets that have extra penetrating power against armor.

deregulation The process of reducing or eliminating regulation, usually in favor of increased private oversight or responsibility.

design margin The added margin of safety built into materials or components to allow for unusual conditions and for the inherent variability in materials.

disposal The permanent removal of radioactive materials from contact with the environment, such as by glassifying and burying them.

dose The quantity of radiation received by an object or person, measured in rads.

dose limit The maximum exposure to radiation permitted by regulation (for example, for nuclear power plant workers).

dose reconstruction The attempt to estimate the dose of radiation that a person has received through exposure to, for example, a nuclear plant accident.

dosimeter A device (such as a film badge) that indicates the total amount of radiation exposure received by a worker in a nuclear power plant. Government regulations specify the maximum exposure that workers are allowed to receive both in a particular amount of time and over their lifetime.

electric charge A fundamental property of matter. Protons have a positive charge, and electrons have a negative charge.

electric force The force exerted by electric charge. Particles with opposite charges are attracted to each other while those with the same charge repel each other. The attraction between protons in the nucleus and the surrounding electrons holds atoms together, but in the tiny space within the nucleus the nuclear force that holds protons together is stronger than the electric force that would otherwise push them apart.

electron The smallest basic particle that makes up the atom. Electrons are negatively charged and orbit the atomic nucleus. The number of and arrangement of electrons in an atom determine its chemical behavior. The flow of electrons is what makes electricity.

element A basic substance such as hydrogen, oxygen, or uranium that cannot be chemically broken down into simpler substances. (An element can be changed into another element only through a nuclear reaction.)

emergency core cooling system A safety system that is designed to prevent the core of a reactor from melting when there is a loss of coolant accident (LOCA). It usually consists of a separate emergency water source, pumps, and connecting pipes.

emission standards Regulations that limit the amount of a substance that can be discharged into the environment.

energy The capacity to do work. Energy can be converted from one form to another. A nuclear power plant uses the heat from nuclear reactions to create steam pressure that turns a turbine whose mechanical energy is then converted to electrical energy.

energy policy Public policies that determine how a nation will assess and develop its energy sources.

energy security The goal of ensuring that a nation has uninterrupted access to a sufficient supply of energy at a reasonable cost. Promoting nuclear power is seen by advocates as an important way to provide energy security by reducing reliance on imported oil.

enrichment The process by which the proportion or concentration of a desired isotope of an element is increased. For example, most nuclear power plants use a fuel in which the proportion of uranium-235 has been enriched from its natural concentration of 1 percent to about 3 percent.

Entomb Nuclear Regulatory Commission term for sealing a nuclear plant as an alternative to decontamination.

entombment The sealing of a decommissioned or accidentally damaged nuclear power plant in concrete.

environment The system of physical, chemical, and biological factors in which all organisms including human beings exist.

environmental impact The effect that a given facility or its activities has on the surrounding environment, such as an increase in chemical pollution or radiation.

environmental impact statement A legally required analysis of the impact (both short-term and long-term) that a proposed development such as a power plant will have on the surrounding environment.

exposure Ionization produced by X-ray or gamma radiation. Exposure can be acute (intense, short-term) or chronic (low-level, of long duration).

external radiation dose Radiation dose received from materials outside the body, mainly gamma rays.

fast neutrons High-energy neutrons such as those ejected by fission reactions.

fertile material Material made of atoms that readily absorb neutrons and become fissionable.

fission The splitting of a heavy nucleus into two roughly equal parts, yielding a large amount of energy and releasing one or more neutrons that can, under the right circumstances, strike other atoms and split them in turn. Fission can occasionally occur spontaneously, but it is usually the result of a nucleus being struck by gamma rays, neutrons, or other particles.

fission products The atoms formed by the pieces into which a heavy nucleus breaks when it fissions. These atoms are usually unstable (radioactive).

fissionable (or fissile) material Isotopes that readily fission when struck by a neutron. Uranium-235 and plutonium-239 are the two most commonly used fissionable materials in nuclear reactors.

fossil fuels Fuels such as coal, oil, or natural gas that are the remnants of ancient biological activity. Fossil fuels have limited supply, and all produce pollution to some degree.

fuel The fissionable material that is used to create the chain reaction in a nuclear reactor.

fuel cycle The entire process by which fuel for nuclear reactors is obtained and used. The cycle includes mining uranium ore; extracting, refining, and enriching the uranium; fabricating the fuel rods; and sometimes, reprocessing the spent fuel. The fuel cycle is complicated by the fact that reactor operation also produces new fissionable materials (such as plutonium) that can be processed into new fuel.

fuel pellet A small cylinder of nuclear fuel (usually uranium oxide), about one-quarter inch thick and one-half inch long.

fuel rods A rod about 12–14 feet long that contains fuel pellets. The rods are enclosed in bundles and inserted into the reactor.

fuel storage pond A pond in which spent fuel is kept after it has been removed from a nuclear reactor. Such fuel is highly radioactive. The water serves as both a coolant and a shield against radiation getting into the environment. Once the radiation level of the fuel has declined sufficiently, it can be shipped to a permanent waste disposal facility. The failure to approve and build such facilities has resulted in a large amount of spent fuel remaining in storage ponds.

fusion The process in which light atomic nuclei come together at high speed and combine into heavier atoms, releasing a large amount of energy. For example, two hydrogen atoms can be fused into a single helium atom.

fusion power The use of controlled fusion reactions to create a steady supply of power. The technology needed to safely contain and maintain a fusion reaction so that it produces more power than it consumes has not yet been developed.

gamma ray High-energy radiation with a short wavelength, similar to X-rays. Gamma rays are highly penetrating but can be stopped by a sufficient thickness of lead.

gas-cooled reactor A reactor in which a gas such as carbon dioxide, rather than water, is used as the coolant.

gaseous diffusion The process by which lighter uranium-235 is separated from heavier uranium-238 in uranium hexafluoride gas because the lighter atoms move more quickly through a membrane.

geiger counter An instrument that measures radioactivity. Radiation ionizes the gas in the tube, allowing it to conduct electricity, which is then measured and correlated to the radiation level.

generation time The time between the ejection of a neutron by a fission event and its causing fission in another nucleus.

glassifying *see* ***vitrification***

global warming The hypothesis that human activities are leading to a gradual increase in the world's average temperature, which may have serious effects on coastal cities, food production, wildlife, and other areas.

gray A unit of absorbed radiation equal to 100 rads.

greenhouse effect The trapping of solar radiation by gases such as carbon dioxide, leading to warming.

half-life The number of years required for a sample of a radioactive substance to lose half of its activity through decay. For example, it takes 24,000 years for a sample of plutonium to lose half of its radioactivity.

heat energy Energy that is transmitted through molecular motion (heat), measured as temperature.

heavy metals Metals that have high atomic weights (usually over 40).

heavy water Water in which the hydrogen atoms consist of deuterium (containing a neutron as well as the usual proton). Heavy water can be used as a moderator in nuclear reactors.

high-level waste Highly radioactive material that remains after spent nuclear fuel is reprocessed. It must be solidified and stored for a long time in a secure facility.

high-temperature gas-cooled reactor A nuclear reactor that is cooled by helium.

highly enriched uranium (HEU) Uranium with a high concentration of uranium-235; it is used for nuclear weapons and in some propulsion reactors in naval vessels.

hot A colloquial term meaning highly radioactive.

integrated waste management system A waste management system that is designed as a single, coordinated set of equipment and procedures for the safe processing and storage of nuclear wastes.

interim storage A temporary facility for storing nuclear wastes until a permanent storage site is available. It is often on the grounds of the nuclear plant itself.

intermediate-level waste Waste that is too radioactive to be considered low-level waste but does not produce significant heat from nuclear decay as does high-level waste.

ion An atom or molecule that has lost or gained one or more electrons. Losing electrons makes a positive ion, and gaining electrons makes a negative ion.

ionization The process by which an atom has electrons stripped away, leaving a positively charged particle.

ionizing radiation Radiation energetic enough to ionize atoms, such as gamma or X-rays (but not visible light or microwaves). Such radiation can disrupt cells in organisms, cause cancer, or damage genetic material.

isotopes Atoms of a given element that differ in atomic weight. For example, uranium has isotopes with weights of 233, 235, and 238. Isotopes can differ in their nuclear properties, such as their susceptibility to fission.

kiloton Measure of explosive force in nuclear weapons. Equivalent to one thousand tons of TNT.

kilowatt One thousand watts. A measurement of the rate of generation or consumption of electricity.

kilowatt-hour The amount of energy produced or consumed by a flow of one thousand watts of electricity for one hour.

leaching The dissolving of a substance through permeable materials, such as a chemical percolating through a layer of soil or a body of water. Leaching can cause wastes or contaminants to diffuse into the environment.

liability Legal responsibility for harm done to others. Can be insured against, but has been legally limited in the case of nuclear power plants.

light-water reactor The most common class of nuclear reactor, which uses ordinary water (as opposed to heavy water, or deuterium) as its coolant. The two most common types of light-water reactor are the boiling-water reactor and the pressurized-water reactor.

linear, no-threshold The theory that risk from radiation is proportional to dose received and that any dose, however small, introduces additional risk.

liquid-metal fast-breeder reactor A nuclear breeder (fuel-producing) reactor that is cooled by the circulation of a liquid metal, usually sodium.

load The total demand for electric power from a facility at a given time.

loss of coolant accident (LOCA) An accident (such as a pipe rupture, blockage, or pump failure) that causes a reactor to lose coolant. This can lead to an increase in core temperature and possibly a meltdown.

low-level waste Waste that contains only a small amount of radioactivity, such as discarded protective clothing. Such waste can usually be buried with minimal containment.

meltdown Overheating of a reactor core (such as through loss of coolant) to the point where it melts and sinks into the earth. This can result in a steam explosion that releases large quantities of highly radioactive material into the atmosphere.

metric ton One thousand kilograms, or about 2,204 pounds. Often used in scientific measurements instead of ordinary "short" tons (two thousand pounds).

mill tailings The slightly radioactive material left after uranium ore is refined to extract uranium oxide.

millirem A dosage of radiation equal to a thousandth of a rem. An average person receives about 100–200 millirem a year from natural and human-made radiation sources; government guidelines recommend that total exposure not exceed 500 millirem.

mixed-oxide fuel Reactor fuel made by combining standard uranium oxide fuel with plutonium that has been obtained from dismantled nuclear weapons or extracted from spent reactor fuel.

moderator A substance that is used to slow down the neutrons produced by fission so that they are more likely to be absorbed by other atoms and perpetuate the chain reaction. Water is the most commonly used moderator; in some reactor designs the same body of water serves as both moderator and coolant. Some reactors use graphite as a moderator.

molten salt breeder reactor A breeder reactor that uses a molten uranium-thorium salt mixture as both fuel and coolant and produces more uranium than it consumes.

monitored retrievable storage The temporary storage of radioactive materials in such a way that it can be monitored for leaks, overheating, or other problems and can be retrieved later for shipment to a permanent storage facility. Spent reactor fuel has been kept in this form of storage in the hope that a permanent storage facility will someday be completed.

multiplication factor The average number of neutrons from each fission reaction that go on to cause another fission. For steady operation, a reactor must have a multiplication factor of exactly one.

mutation A genetic effect brought about by alteration of DNA, such as through exposure to natural or artificial radiation.

neutron An atomic particle that has the same mass as a proton but has no electric charge. Protons and neutrons together make up the atomic nucleus. Because neutrons have no charge, they can approach (and possibly split) a nucleus without having to overcome electrical repulsion.

neutron poison A material that absorbs neutrons. If such a material accumulates in a nuclear reactor it can halt the chain reaction.

nuclear energy Energy released by changes in an atomic nucleus, such as by its splitting (fission) or combining (fusion).

nuclear island The nuclear reactor complex as a whole, including containment building, auxiliary cooling system, and other apparatus.

nuclear pile A relatively simple, early design of reactor made of bricks containing uranium and a moderating material such as graphite.

nuclear steam supply system A nuclear reactor and steam-generation system that provides steam that is used in a turbine to generate electricity.

nuclear weapon A device that uses an arrangement of fissionable material to produce an extremely violent, uncontrolled nuclear chain reaction that creates a very powerful explosion. For greater power, the explosion can be used to create a fusion (thermonuclear) reaction.

nuclear weapons capability The ability of a nation or nonnational group to design or build nuclear weapons in a relatively short period of time. Capability requires expertise, suitable equipment, and access to fissionable materials.

nucleon A proton or neutron, normally found in the nucleus of an atom.

nuclide A type of atom, specified by atomic mass, atomic number, and energy state, that has particular nuclear properties (including radioactive intensity and half-life).

once-through fuel cycle A simple fuel cycle in which nuclear fuel is used only once, with the spent fuel being stored as radioactive waste rather than being reprocessed to extract fissionable uranium-235 or plutonium-239.

operating costs The cost of routine operation of a nuclear power plant.

operating license The legal document that permits a nuclear power plant to be operated for commercial power generation.

pathway analysis Analysis of the routes or processes by which radioactive or toxic material can reach human beings.

peak load The maximum level of demand for electricity from a power plant measured over a period of time such as a day, season, or year.

percolation The downward movement of water (and whatever is dissolved in the water) through permeable layers of rock.

photon The fundamental unit (quantum) of electromagnetic radiation, having a characteristic energy level ranging from radio to light to X-rays and gamma rays.

plutonium An element that is artificially produced by bombarding uranium with neutrons in the core of a reactor. Plutonium in turn is able to fission easily and can be used as reactor fuel or for nuclear weapons.

pollution In general, the deliberate or accidental introduction of harmful substances, heat, or radiation into the environment (ground, water, or atmosphere).

power plant Any facility that generates electricity from some form of energy. A nuclear power plant uses the heat from splitting atoms (fission) to generate steam that turns a turbine to generate electricity.

pressure vessel The steel enclosure that surrounds the core of a reactor. It is designed to withstand high pressures and temperatures in order to prevent radioactive material from escaping in the event of an accident involving the core.

pressure-tube reactor A nuclear reactor in which fuel is placed inside a large number of tubes through which is circulated a high-pressure coolant.

pressurized water reactor A nuclear reactor that is cooled by water that is kept under high pressure to prevent it from boiling, allowing it to carry a large amount of heat. The coolant then heats the water in a separate loop from which steam for the turbine is produced.

primary coolant The liquid or gas that provides most of the cooling for the fuel elements in the reactor core.

primary loop The part of a reactor's cooling system that sends coolant through the reactor core.

proliferation The spread of nuclear weapons, nuclear weapons capability, or weapons components (including fissionable material) to nations not currently having nuclear capability.

proton A positively charged particle that together with the neutron makes up the nucleus of the atom.

public policy Plans and procedures established by a government agency to deal with issues of public concern, such as the development of nuclear power, the storage of nuclear waste, and the prevention of nuclear proliferation.

rad (radiation absorbed dose) A measure of the amount of ionizing radiation that has been absorbed by any material, including human tissue.

radiation Particles or electromagnetic waves that go out from a source into surrounding space. The disintegration of an unstable atom produces radiation.

radiation sickness Illness caused by exposure to a large dose of radiation (100 rem or more). Symptoms include nausea, vomiting, intestinal bleeding, and anemia.

radiation therapy Use of radiation in medicine, such as to kill some types of tumors.

radioactive tracer Small amount of radioactive material that can be used to track the path of a substance through the body.

radioactive waste Unwanted radioactive materials produced by a nuclear facility, or ordinary materials that have been mixed with radioactive substances or made radioactive through exposure to gamma rays or other radiation.

radioactivity The spontaneous emission of alpha or beta particles (or gamma rays) by atoms of unstable elements such as uranium.

radioisotope An unstable isotope of an element, which will eventually undergo radioactive decay.

radionuclide Any radioactive isotope.

radon A radioactive gas that is naturally produced through decay of a radium decay product in the earth. It can accumulate in unventilated building spaces (such as basements); prolonged breathing of radon can induce lung cancer. Radon represents more than half of the average background radiation dose.

reactor A device that contains a controlled nuclear fission chain reaction. The reaction can be used to generate electricity, carry on nuclear research, or produce plutonium for nuclear weapons.

reactor vessel The container of a reactor's nuclear core. It can be a pressure vessel, a low-pressure vessel, or a prestressed concrete enclosure.

reactor year The operation of a nuclear reactor for one year. Often used in statistics such as accidents per reactor year or expected reactor years between major accidents.

recycling The extraction and reuse of remaining fissionable material in spent fuel.

refueling The removal of spent fuel from a reactor and its replacement with new fuel.

regulatory policy Policies that specify each aspect of a nuclear plant's operations, including training requirements, safety standards, reporting requirements, and releases into the environment.

rem Acronym for "roentgen equivalent man," a unit of radiation effect on the human body.

renewable energy sources Sources of energy that last indefinitely without being depleted, such as solar and hydroelectric power. Uranium is not fully renewable as an energy source, but its life can be extended through use of breeder reactors.

repair enzymes Substances in the body that may be able to repair damage to DNA caused by low doses of radiation.

repository A permanent storage facility for nuclear wastes, including spent fuel.

reprocessing The chemical treatment of spent nuclear fuel to remove remaining uranium and plutonium from the fission products, allowing it to be used to make new fuel.

roentgen The unit of exposure to gamma rays or X-rays, named for Wilhelm Roentgen, who discovered X-rays in 1895.

safeguards Regulations and procedures for monitoring and controlling the storage and shipping of nuclear materials to prevent their being diverted by unauthorized persons or groups, including terrorists and nations seeking nuclear weapons.

safety system In general, a combination of electronic, mechanical, and instrumentation systems designed to detect or prevent accidents that might endanger the reactor or the surrounding community.

Safstor Nuclear Regulatory Commission term for decommissioning a nuclear plant by "mothballing" it—letting it run down gradually until it has lost most of its radioactivity.

scram A rapid shutdown of a reactor core by inserting neutron absorbers (usually rods) when there is danger of the reaction getting out of control. Since this can happen very quickly, it is usually an automatic process triggered by specified conditions rather than being triggered by the operator.

secondary loop The part of a nuclear power plant that is not in contact with the core, but uses heat carried from the core to generate steam to produce electricity.

shielding Materials such as water, lead, or concrete that are placed around radioactive material to protect personnel and the environment from radiation exposure.

shutdown Any halt in operation of a nuclear power plant.

slow neutrons Neutrons that are slower than those ejected by a fission event. Slow neutrons are much more effective than fast neutrons for causing fission in uranium-235, so reactors use a moderator to slow down the neutrons produced by fission.

spent fuel Nuclear fuel whose fissionable material has been depleted to the point that it can no longer sufficiently sustain a chain reaction. Spent fuel is both thermally hot and highly radioactive and must be handled with extreme caution.

spent fuel pool A deep pool of water near a nuclear reactor in which spent fuel can be stored until it cools enough to allow it to be moved to a more permanent storage facility.

stable nucleus A nucleus that can exist indefinitely without spontaneously coming apart; it is thus nonradioactive.

steam generator A device that uses the heat from pressurized water or hot gas to create steam for use in power generation.

storage The process of isolating, containing, and potentially recovering radioactive materials. Storage facilities must be designed to be monitored and maintained for a specific period of time.

thermal pollution The discharge of excess heat, such as into rivers or lakes. This can harm living things and disrupt the ecosystem.

thermal reactor A nuclear reactor where the chain reaction is sustained mainly by thermal neutrons.

thorium-232 A natural isotope that can be bombarded to create the fissionable isotope uranium-233.

threshold theory The idea that there is some level of total radiation exposure below which there are no health effects. Most scientists now believe there is no such threshold, although effects of low doses of radiation are hard to prove.

tokamak A donut-shaped device used to contain a nuclear fusion reaction.

transuranic wastes Nuclear wastes that consist of heavy elements that are derived from uranium, isotopes of plutonium, and other elements with atomic numbers greater than 92 (uranium). Such wastes are produced mainly from reprocessing spent fuel and from use of plutonium in building nuclear weapons.

turbine An engine that consists of blades attached to a rotating shaft. It can be used to create the kinetic energy from steam to mechanical motion that in turn can be used to generate electricity.

uranium A natural radioactive element that is heavy, hard, shiny, and metallic. Certain isotopes of uranium are suitable for use in nuclear reactors.

uranium milling The process of extracting uranium from ore.

uranium-233 A fissionable uranium isotope suitable for use in making nuclear weapons.

uranium-235 A fissionable uranium isotope that makes up less than 1 percent of all natural uranium. It is the most common fuel used in nuclear power plants.

uranium-238 The most common (99 percent) isotope of uranium. It is not very fissionable but can be turned into the fissionable plutonium-239 in a nuclear reactor.

vitrification A process that turns radioactive wastes into a glasslike substance that is relatively inert and easy to store.

waste disposal system The overall system of processing facilities, storage sites, and transportation systems for handling radioactive wastes.

water table The upper boundary of the area beneath the ground surface that is completely saturated with water.

watt A unit of power (the rate at which work is done). For electricity, it is a measure of the rate of generation or consumption.

watt-hour The amount of energy involved in generating or consuming one watt of electricity for one hour.

weapons-grade uranium Highly enriched uranium (from 20 to 70 percent uranium-235) that can be used to make nuclear weapons. (Commercial enriched uranium for power plants is only about 3.5 percent uranium-235.)

yellowcake A uranium-rich yellow powder that is the residue left after pouring crushed uranium ore into an acid that dissolves the uranium, allowing the acid solution to evaporate.

canister The outer metal container in which glassified high-level radioactive waste or spent fuel rods are placed.

cask A container used to transport and shield waste canisters or other radioactive materials during shipping.

chain reaction The process by which one nuclear fission emits neutrons that in turn cause other fissions, resulting in a self-sustaining release of nuclear energy.

chemical reaction A process in which one or more chemical substances interact, creating different substances and using or releasing energy. Unlike nuclear reactions, chemical reactions do not affect the cores of atoms and do not change one element into another.

cladding Material (such as steel or a metal alloy) that is used to encapsulate and shield pieces of radioactive fuel in a reactor.

cold fusion Hypothetical fusion reaction achieved by infiltrating atoms into a tightly packed crystal lattice rather than through the use of high temperature and pressure. Failure to replicate the experiments of the late 1980s has led most scientists to reject the existence of cold fusion.

colloid Small particles (10^{-9} to 10^{-6} meters in size) that are suspended in a solvent such as water. Natural colloids arise from clay minerals in contact with water, but radioactive waste can also escape into the environment in this form.

commissioning The process of making a nuclear power plant operational, which involves a lengthy testing and regulatory process.

condenser The pipes and other apparatus that allow steam to cool and turn back into water.

containment building The thick, steel-reinforced concrete building that surrounds the pressure vessel of a nuclear reactor. It represents the final barrier to the release of radioactive materials into the atmosphere.

containment system The layers of packaging and other systems designed to prevent radioactive materials from escaping into the environment during shipping.

contamination Radioactive material that has gotten into an undesired location, such as through leaks, leaching, or some other process.

control rods Long, thin rods made of a material that absorbs neutrons. Inserting rods into the reactor reduces the number of neutrons that can cause fission and thus brings the nuclear chain reaction to a halt.

control room The room in a nuclear power plant from which all operations are monitored and controlled through a large array of gauges, indicators, and controls.

coolant A fluid (usually water) that is circulated through the core of a nuclear reactor to remove the heat generated by the fission process. Some reactors use liquid metal (sodium) or a gas as a coolant.

cooling pond A pool or body of water used to dissipate heat from power plant operations through evaporation.

cooling system The system of pumps, pipes, and so on that carries coolant through a nuclear power plant.

cooling tower A tall, usually funnel-shaped structure that removes heat from water that has condensed from steam in the condenser. Cooling the water before putting it back in the environment prevents damage to wildlife.

core The central part of a nuclear reactor where the fission chain reaction takes place. It contains fuel rods, a moderator, and control rods. A coolant is continuously circulated through the core to remove heat, which is used to generate electric power.

cosmic rays High-energy particles from space that make up much of people's background radiation exposure.

critical The condition where a nuclear reaction is self-sustaining because the neutrons released by each fission can (on the average) trigger one new fission.

critical mass The minimum amount of fissionable material that for a given geometric arrangement allows for a self-sustaining chain reaction.

curie A unit of radioactive intensity equivalent to a gram of radium, representing thirty-seven billion atomic disintegrations per second.

decay The disintegration of an unstable atomic nucleus, resulting in the emission of charged particles and/or photons (gamma rays).

decommission Removal from service of a nuclear power plant that has ended its service life. Includes the dismantling and safe disposal of the reactor vessel and the reduction of radioactivity at the site so that it is available for unrestricted use.

Decon Nuclear Regulatory Commission term for decommissioning a nuclear plant by promptly decontaminating it and restoring the site to normal use.

decontamination The process of removing radioactive materials that have become mixed with the environment, a worker's clothing, or other materials.

defense in depth The design philosophy for nuclear facilities in which several different barriers are placed, each believed sufficient to prevent an escape of radioactive materials. For example, American reactors have both an inner containment and a strong outer building.

depleted uranium Uranium left over from the enrichment process that segregates and concentrates uranium-235. The depleted uranium is thus much lower in uranium-235 than natural unprocessed uranium. Because of its density, depleted uranium is used to make shells or bullets that have extra penetrating power against armor.

deregulation The process of reducing or eliminating regulation, usually in favor of increased private oversight or responsibility.

design margin The added margin of safety built into materials or components to allow for unusual conditions and for the inherent variability in materials.

disposal The permanent removal of radioactive materials from contact with the environment, such as by glassifying and burying them.

dose The quantity of radiation received by an object or person, measured in rads.

dose limit The maximum exposure to radiation permitted by regulation (for example, for nuclear power plant workers).

dose reconstruction The attempt to estimate the dose of radiation that a person has received through exposure to, for example, a nuclear plant accident.

dosimeter A device (such as a film badge) that indicates the total amount of radiation exposure received by a worker in a nuclear power plant. Government regulations specify the maximum exposure that workers are allowed to receive both in a particular amount of time and over their lifetime.

electric charge A fundamental property of matter. Protons have a positive charge, and electrons have a negative charge.

electric force The force exerted by electric charge. Particles with opposite charges are attracted to each other while those with the same charge repel each other. The attraction between protons in the nucleus and the surrounding electrons holds atoms together, but in the tiny space within the nucleus the nuclear force that holds protons together is stronger than the electric force that would otherwise push them apart.

electron The smallest basic particle that makes up the atom. Electrons are negatively charged and orbit the atomic nucleus. The number of and arrangement of electrons in an atom determine its chemical behavior. The flow of electrons is what makes electricity.

element A basic substance such as hydrogen, oxygen, or uranium that cannot be chemically broken down into simpler substances. (An element can be changed into another element only through a nuclear reaction.)

emergency core cooling system A safety system that is designed to prevent the core of a reactor from melting when there is a loss of coolant accident (LOCA). It usually consists of a separate emergency water source, pumps, and connecting pipes.

emission standards Regulations that limit the amount of a substance that can be discharged into the environment.

energy The capacity to do work. Energy can be converted from one form to another. A nuclear power plant uses the heat from nuclear reactions to create steam pressure that turns a turbine whose mechanical energy is then converted to electrical energy.

energy policy Public policies that determine how a nation will assess and develop its energy sources.

energy security The goal of ensuring that a nation has uninterrupted access to a sufficient supply of energy at a reasonable cost. Promoting nuclear power is seen by advocates as an important way to provide energy security by reducing reliance on imported oil.

enrichment The process by which the proportion or concentration of a desired isotope of an element is increased. For example, most nuclear power plants use a fuel in which the proportion of uranium-235 has been enriched from its natural concentration of 1 percent to about 3 percent.

Entomb Nuclear Regulatory Commission term for sealing a nuclear plant as an alternative to decontamination.

entombment The sealing of a decommissioned or accidentally damaged nuclear power plant in concrete.

environment The system of physical, chemical, and biological factors in which all organisms including human beings exist.

environmental impact The effect that a given facility or its activities has on the surrounding environment, such as an increase in chemical pollution or radiation.

environmental impact statement A legally required analysis of the impact (both short-term and long-term) that a proposed development such as a power plant will have on the surrounding environment.

exposure Ionization produced by X-ray or gamma radiation. Exposure can be acute (intense, short-term) or chronic (low-level, of long duration).

external radiation dose Radiation dose received from materials outside the body, mainly gamma rays.

fast neutrons High-energy neutrons such as those ejected by fission reactions.

fertile material Material made of atoms that readily absorb neutrons and become fissionable.

fission The splitting of a heavy nucleus into two roughly equal parts, yielding a large amount of energy and releasing one or more neutrons

A freelance technical writer with twenty years' experience, Harry Henderson has written numerous books on information technology; biography; and history of science, social issues, and other topics for both young people and adults. His works include *Internet How-To* (Waite Group Press, 1994), *Nuclear Physics* (Facts on File, 1998), *Issues in the Information Age* (Lucent Books, 1999), and *Privacy in the Information Age* (Facts on File, 1999). He believes that it is vital that the public have a clear understanding of emerging technologies and the issues and challenges that arise from them. He lives in El Cerrito, California, with his wife, Lisa Yount, who is also a prolific science writer.

Index

Abalone Alliance, 131
Abramson, Barry, on nuclear plants, 26
Acceptable Risks? (video), 202
Accidents, 115, 152, 160
  nuclear power and, 101–102
  preparedness for, 11
  report on, 10
  See also Chernobyl; Three Mile Island
Acheson-Lillienthal report (1946), 105
Activity, defined, 219
Adsorption, defined, 219
Advanced boiling-water reactor (ABWR), construction of, 41
AEC. *See* Atomic Energy Commission
AFL-CIO, challenge by, 36
Allen v. United States, 180
Alliance for Nuclear Accountability (ANA), 131
Alpha particles, 3
  defined, 219
Alternative energy, xi, xii, 156
  introduction of, 154
American Academy of Health Physics, 132
American Environmental Health Studies Project, 132
American Institute of Physics (AIP), 132
American Nuclear Society, 133, 185
  Division of Nuclear Physics, 133
American West, cultural perceptions of, 181
Amory Lovins (video), 205
Annals of Nuclear Energy, 186
Antarctic Treaty (1959), 119
Antinuclear movement, 149, 151, 156, 161, 171, 176, 184
  exposé of, 157–158
  in Mexico, 157
  social values of, 165
  Yucca Mountain and, 105–112
Aquifer, defined, 219
Arkansas Nuclear 1, statistics on, 63
Arkansas Nuclear 2, statistics on, 63
Arthur Anderson, on nuclear utilities, 103
Associated Press, 25, 215
  on nuclear power, 43
Atom, 1, 2, 156
  defined, 219
Atomic Bomb Casualty Commission, 180
Atomic energy. *See* Nuclear energy
Atomic Energy Act (1946), 163
  AEC and, 7, 33, 159–160
  described, 116
Atomic Energy Act (1954), 116–117
  private ownership and, 34
Atomic Energy Commission (AEC), 158, 165
  on accidents, 10
  Atomic Energy Act and, 7, 33, 159–160
  Cooperation Power Demonstration Program and, 34
  demise of, 162
  formation of, 6, 33, 159–160, 163
  Fort Greely and, 185
  history of, 159–160, 163
  mandates of, 11
  nuclear power and, 12
  Reactor Safeguards Committee of, 34
  replacement of, 37, 38
  report by, 36

safety/licensing and, 37
Shippingport and, 166
ten-year plan and, 35
Atomic Energy Commission (France), 133
Atomic Energy Council (Taiwan), 133
Atomic Energy Insights, 186
Atomic mass, defined, 219
Atomic number, defined, 219
Atomic Reclamation and Conversion Project, 134
Atomic Safety and Licensing Appeal Board (AEC), 37
Atomic theory, 1
"Atoms for Peace," 6, 159, 163–164, 174
Australian Nuclear Science and Technology Organization (ANSTO), 134

Babbitt, Bruce, 22
Babcock and Wilcox, nuclear market and, 36
Backfill, 93, 94
defined, 220
Baltimore Gas & Electric Co. v. NRDC (1983), 122
Baltimore Gas and Electric
Calvert Cliffs and, 24
extension for, 42
steam generator replacement by, 26
Ban Waste Coalition (Ward Valley, CA), 134
Barriers
engineered, 93–94, 95
geologic, 89, 95
Baruch, Bernard, biographical sketch of, 45
Baseline information, defined, 220
Base load, defined, 220
Base loaded, defined, 220
Beaver Valley 1, statistics on, 63
Beaver Valley 2, statistics on, 64
Bechtel Group, Inc., contract for, 41
Becquerel, Antoine Henry, 2, 31, 32
Becquerel, defined, 220
Bellona Foundation, 134
Beta particles, 3
defined, 220
Bethe, Hans Albrecht, 155
biographical sketch of, 46
Big Rock Point, statistics on, 64
Bohr, Niels, biographical sketch of, 47
Boiling water reactor (BWR), 35, 41, 59 60–62, 220
Boxer, Barbara, Yucca Mountain and, 110
Braidwood 1, statistics on, 64
Braidwood 2, statistics on, 64
Breeder reactors, 18, 27, 37
call for, 36
defined, 220
British thermal unit (BTU), defined, 220
Brookhaven National Laboratory, 10
Brookhaven Report, 10, 35
Brown's Ferry 1, statistics on, 64
Brown's Ferry 2, statistics on, 64
Brown's Ferry 3, statistics on, 64
Brown's Ferry Nuclear Plant, incident at, 12, 38
Brunswick 1, statistics on, 64
Brunswick 2, statistics on, 65
Bulletin of the Atomic Scientists, The, 153, 186
Bureau of Land Management, 22
Burn-up, defined, 220
Byron 1, statistics on, 65
Byron 2, statistics on, 65

Caldicott, Helen B., biographical sketch of, 47–48
California Energy Commission, 135
Callaway 1, statistics on, 65
Calvert Cliffs 1, statistics on, 65
Calvert Cliffs 2, statistics on, 65
Canadian Nuclear Society, 135
Cancer, 7, 9, 114, 179
dose, 220
Canister, defined, 221
Cantlon, John E., 110
Carnegie Endowment for International Peace, 173, 174
Carolina Environmental Study Group, 38–39
Carter, Jimmy
biographical sketch of, 48
Energy Reorganization Act and, 38
nonmilitary nuclear research and, 39
spent nuclear fuel and, 38
Casks
defined, 221
testing, 21
Catawba 1, statistics on, 65
Catawba 2, statistics on, 65
Cathode rays, 1
Center for Strategic and International Studies (CSIS), 161
Cesium-137, 18
Chadwick, James, neutrons, 3, 32
Chain reaction, defined, 221
Changes (video), 211
Characterization study, 108–109

Chelyabinsk: The Most Contaminated Spot on the Planet (video), 211
Chemical reaction, defined, 221
Chernobyl, 25, 147, 151, 177–178
accident at, xi, 4, 14–17, 102, 114
cleanup at, 171
glasnost and, 16
legacy of, 16–17, 23
shutdown at, 43
Chernobyl: Chronicle of Difficult Weeks (video), 211
Citizens Clearinghouse for Hazardous Wastes, 135
City of West Chicago v. Kerr-McGee (1982), 122
Civilian Radioactive Waste Management Office, 40
Cladding, defined, 221
Clinch River breeder reactor, 38, 40
Clinton 1, statistics on, 65
Clinton, Bill
waste storage and, 20
WIPP and, 42
Yucca Mountain and, 44
Cohen, Bernard, 155
biographical sketch of, 48–49
Cold fusion, defined, 221
Cold War, 22
nuclear industry and, 156
Colloid, defined, 221
Comanche Peak 1, statistics on, 66
Comanche Peak 2, statistics on, 66
Combustion, 1
Combustion Engineering, nuclear market and, 36
Commissioning, defined, 221
Committee for Nuclear Responsibility, 136
Committee for Review and Evaluation of the Medical Use Program (NRC), 162
Commoner, Barry, biographical sketch of, 49
Commonwealth Edison Co. v. NRC (1987), 122–123
Commonwealth Edison, Dresden 1 and, 35
Condenser, defined, 221
Containment building, defined, 221
Containment system, defined, 221
Contamination, 114
defined, 221
Control rods
defined, 221
problems with, 15
Control room, defined, 221
Coolant, defined, 222
Cooling pond, defined, 222
Cooling system, 10
defined, 222
Cooling tower, defined, 222
Cooper 1, statistics on, 66
Cooperation, international, 119–121
Core, defined, 222
Cosmic rays, defined, 222
County of Suffolk v. Long Island Lighting Company (1984), 123
Court cases, 122–130
Critical, 115
defined, 222
Critical mass, defined, 222
Critical Mass Energy Project v. NRC (1987), 123
Critical Mass, Voices for a Nuclear-Free Future (video), 206
Criticality
external, 101
waste burial and, 100, 101
Yucca Mountain and, 108
Crowley, Kevin D., 183
Crystal River 3, statistics on, 66
Culture of safety, 176
Culture, nuclear power and, 170
Curie, defined, 222
Curie, Marie Sklodowska, radioactivity and, 2, 31–32
Curie, Pierre, radioactivity and, 2, 31–32
Current Bibliography on Science and Technology: Nuclear Engineering, 186

Dalton, John, atomic theory and, 1
Davies, John B., on repositories/water, 107
Davis Besse 1, statistics on, 66
Deaf Smith, 88
Deafsmith: A Nuclear Folktale (video), 213
Death Valley National Park, Yucca Mountain and, 110
Decay, defined, 222
Decommission, 24, 27, 39, 159
defined, 222
Decon, 23
defined, 222
Decontamination, 24, 95
defined, 222
Defense in depth, defined, 222
Department of Defense, regulation by, 121
Department of Energy (DOE), 143, 182
accident rates and, 115
Bechtel and, 41

characterization study and, 108–109
Civilian Radioactive Waste Management Office and, 40
creation of, 38
disposal issues and, 43, 88
GAO and, 107, 110
nuclear construction costs and, 103
nuclear energy and, 28–29
problems at, 110
radioisotopes and, 184–185
reference design and, 96
regulation by, 39, 40, 110, 121
spent nuclear fuel and, 87
storage and, 108–109, 169
surplus plutonium and, 43
transport issue and, 114
waste package design and, 94, 97
Yucca Mountain and, 95, 106, 109, 115
Department of Transportation, regulation by, 121
Deregulation, 24, 104, 162, 164
defined, 223
Design margin, defined, 223
Diablo Canyon, 7
Diablo Canyon 1, statistics on, 66
Diablo Canyon 2, statistics on, 66
Dietz, David, on nuclear power, 5–6
Disposal, xiii, 43, 80, 85, 87, 88, 91, 106–112, 154, 181, 183
defined, 223
direct, 183
geologic, 89–90
safe, 95–96
DOE. *See* Department of Energy
Domenici, Pete, on nuclear technology, 29
Donald C. Cook 1, statistics on, 66
Donald C. Cook 2, statistics on, 67
Door into the New World: Nuclear Energy and American Society, A (video), 202
Dose, 99
absorbed, 219
background, 220
defined, 223
effect and, 9
external, 224
limit, 223
reconstruction, 223
Dosimeter, defined, 223
Dresden 1, 35
Dresden 2
statistics on, 67
steam pipe break for, 37
Dresden 3, statistics on, 67
Duane Arnold, statistics on, 67
Duke Power, extension for, 42
Duke Power Co. v. Carolina Env. Study Group (1978), 123–124

Earth Day (1970), xi
Earth Summit (Kyoto), 28, 41–42
Earthquakes
waste burial and, 100
Yucca Mountain and, 106, 107
Ecologia, 136
Eddleman v. NRC (1987), 124
Einstein, Albert, 159
atomic bomb and, 33
Roosevelt and, 5, 33
theory of relativity and, 4, 32
Eisenhower, Dwight D.
AEC and, 35
"Atoms for Peace" speech and, 6, 34
biographical sketch of, 49–50
Price-Anderson Act and, 35
Shippingport and, 7
Eizenstat, Stuart, on nuclear energy, 28
Electric charge, defined, 223
Electric force, defined, 223
Electric Power Research Institute, 136
Electricity, xi, 2, 29, 104, 165, 170
from nuclear power, 79 (table)
oil embargo and, 27
U.S. (by energy source), 62 (fig.)
Electron, defined, 223
Element, defined, 223
Emergency core cooling system, defined, 223
Emission standards, defined, 224
Emplacements, 93, 94
End of Nuclear Power, The (video), 206
Energy and Environmental Programme, 173
Energy crisis, 26, 37, 170
electricity and, 27
Energy
defined, 224
economics of, 25
See also Nuclear energy
Energy Information Administration (EIA) (DOE), 27, 42, 143
Energy Policy Act (1992), 89, 181
Energy policy, defined, 224
Energy Reorganization Act (1974), 38, 118
Energy Reorganization Act (1977), 38
Energy Research and Development Administration (ERDA), 37
establishment of, 38

Energy security, defined, 224
English v. General Electric Co. (1990), 124–125
Enrichment, 18, 81, 82, 181, 183, 226
defined, 224
Entomb, defined, 23, 224
Entombment, defined, 224
Environment, 7–9, 29, 101, 162, 172, 179, 183
defined, 224
nuclear power and, 37, 170
Environmental impact, 154
defined, 224
Environmental impact statements, 37, 111
defined, 224
Environmental Protection Agency (EPA)
regulation by, 89, 121
repositories and, 89
storage and, 169
Environmental Protection Agency Radiation Protection Program, 144
EPA. *See* Environmental Protection Agency
Era of Atomic Energy, 6
Eskom, 43
EURATOM Cooperation Act (1958), 119–120
European Nuclear Society, 137
Experimental Breeder Reactor 1, electricity from, 34
Exposure, defined, 224

Fast breeder nuclear reactor, 43
Fast neutrons, defined, 224
Federal Emergency Management Agency (FEMA), report by, 40
Federal Energy Administration, creation of, 38
Federal Trade Commission, nuclear industry ads and, 43
Feinstein, Dianne, Yucca Mountain and, 110
Fermi 1
decommissioning of, 159
partial meltdown for, 36
Fermi 2, statistics on, 67
Fermi, Enrico
biographical sketch of, 50
fission and, 33
Manhattan Project and, 33
neptunium and, 32
radioactive isotopes and, 4
Fertile material, defined, 224
Fighting for Safe Energy (video), 206
First Conference on the Peaceful Uses of Atomic Energy (1955), 172
Fission, 3–4, 28, 32, 33, 81, 83, 115, 154, 179
controlled, 5
defined, 224
discovery of, xii
products, 225
self-sustaining, 5
Fissionable/fissile material, 5
defined, 225
Fitting the Pieces: Managing Nuclear Waste (video), 213
Food and Drug Administration (FDA), regulation by, 121
Forbes, on nuclear power program, 103
Ford, Daniel, 38
on decommissioning, 24
Ford, Gerald, Energy Reorganization Act and, 38
Foreign policy, nuclear power and, 171
Fort Calhoun 1, statistics on, 67
Fort Greely, 185
Fossil fuel, 1, 154, 159
defined, 225
reducing use of, 42, 158
Friends of the Earth, report by, 38
Frisch, Otto, 4, 32
Fuel
defined, 225
fabrication, 82
See also Fossil fuel; Mixed- oxide (mox) fuel; Spent fuel
Fuel cycle, 79–85, 87–102, 152, 154
defined, 225
once-through, 229
Fuel pellets, 17, 18, 82, 83
defined, 225
Fuel rods, defined, 225
Fuel storage pond, defined, 225
Fusion, 28, 152, 154, 183
defined, 225
Fusion Power Associates, 137
Fusion power, defined, 225

Gamma rays
defined, 225
discovery of, 32
GAO. *See* General Accounting Office
Gas centrifugation, 82
Gas-cooled reactor, defined, 226
Gaseous diffusion, 82
defined, 226
Geiger counter, defined, 226
General Accounting Office (GAO)
DOE data and, 107, 110
Yucca Mountain and, 111

General Electric
BWR by, 35, 41
Dresden 2 reactor and, 37
nuclear market and, 36
reactor sales by, 105
Generation time, defined, 226
Geothermal power, 154
German nuclear power plants phase-out, 26–27, 44
Ghost Dance Fault, 106
Glassifying. *See* Vitrification
Global warming, 162, 173
concern about, 28
defined, 226
nuclear power and, 104–105
Gofman, John, 155, 178
biographical sketch of, 50–51
Grand Gulf 1, statistics on, 67
Gray, defined, 226
Greenhouse effect, xii, 28, 159
defined, 226
reducing, 41–42, 104
Greenpeace International, 137
Greenpeace (U.S. office), 137
Groundwater
concerns about, 91, 98, 182
contamination of, 107, 108
flow of, 99, 110

H. B. Robinson 2, statistics on, 67
Haddam Neck, statistics on, 68
Hahn, Otto, 4, 32
Half Lives (video), 203
Half-life, defined, 226
Hall, Joel T., 109
Hanford Education Action League (HEAL), 138
Hanford Nuclear Reservation, 88, 178
environmental/health issues at, 179
Hatch 2, statistics on, 68
Haves and Have-Nots (video), 203
Health issues, 7–9, 172, 179
Health Physics Society, 138
Heat energy, defined, 226
Heavy metals, defined, 226
Heavy water, defined, 226
Hickenlooper, Bourke B., biographical sketch of, 51
High-level waste, 19, 84, 114
defined, 226
legacy of, 87
reprocessing, 22
storing, 40, 87
Highly enriched uranium (HEU), 183
defined, 226
High-temperature gas-cooled reactor, defined, 226
Hiroshima, 5, 33
Hitachi, ABWR by, 41
Hitler, Adolf, 5
Hope Creek 1, statistics on, 68
Hot, defined, 226
Human intrusion, waste burial and, 100–101, 108
Humboldt Bay, closing, 25
Hydroelectric power, 28, 154

Incident at Brown's Ferry (video), 211
Indian Point 2
statistics on, 68
steam tube rupture at, 44
Indian Point 3, statistics on, 68
Industrial Advisory Group (AEC), 6, 34
Institute for Energy and Environmental Research, 138
Institute of Nuclear Power Operations (INPO), 139, 176
Integrated waste-management system, defined, 226
Interagency Review Group on Nuclear Waste Management, Report to the President by, 90
Intermediate-level waste, defined, 227
International Atomic Energy Agency (IAEA), 35, 139, 154
International Commission on Radiation Protection, 41
International House of Japan, 173
International Nuclear Safety Center (INSC), 139, 167–168
Internet, 29
nuclear issues and, xiii
Iodine, 98
Iodine–129, 99
Ion, defined, 227
Ionization, defined, 227
Ionizing radiation, defined, 227
Isotopes, 4, 98
defined, 227
fissionable, 81

Jackson, Henry M., biographical sketch of, 51
James Fitzpatrick 1, statistics on, 68
Jersey Central Power and Light Company, 36
Johnson, Lyndon B.
NEPA and, 37
Private Ownership of Special Nuclear Materials Act and, 36
Joint Committee on Atomic Energy (JCAE), 33, 34
Joseph M. Farley 1, statistics on, 68

Joseph M. Farley 2, statistics on, 69
Journal of Nuclear Science and Technology, 187
Judith Johnsrud and Dr. Donnell W. Boardman (video), 207

Kendall, Henry, 38
Kennedy, John F., nuclear power and, 36
Kerr-McGee v. Farley (1997), 125
Kewaunee, statistics on, 69
Kiloton, defined, 227
Kilowatt, defined, 227
Kilowatt-hour, defined, 227
Klaproth, Martin Heinrich, uranium and, 2
Kyoto environmental summit, 28, 41–42

Laguna Verde nuclear power plant, opposition to, 157
Lasalle 1, statistics on, 69
Lasalle 2, statistics on, 69
Lawrence, Ernest O., Manhattan and, 33
Leaching, defined, 227
League of Women Voters, 25
Levenson, Milton, 183
Liability, defined, 227
Life, Death, and the Nuclear Establishment (video), 207
Light-water reactors, 102
  defined, 227
Lilienthal, David E., biographical sketch of, 52
Limerick 1, statistics on, 69
Limerick 2, statistics on, 69
Linear, no threshold, defined, 227
Liquid-metal fast breeder reactor, defined, 227
Load, defined, 227
LOCA. *See* Loss of coolant accident
LOFT. *See* Loss of Fluid Test Reactor
Long Island Lighting Company, 166
Los Alamos National Laboratory, 20, 33
  study by, 107
  wildfire at, 44
  Yucca Mountain and, 108
Loss of Fluid Test Reactor (LOFT), 11
Loss-of-coolant accident (LOCA), 10–11, 13
  defined, 228
Low Level Waste Policy Act (1980), 21, 118–119
Low-level waste, 84
  defined, 228
  minimizing, 21
  shipping, 21
  storing, 21–22

MacKenzie, James, 159
Madres Veracruzanas, 157
Maine Yankee, statistics on, 69
Manhattan Project, 22, 33, 160, 180
Martin, McChesney, on fossil fuel pollution, 7
McGuire 1, statistics on, 69
McGuire 2, statistics on, 70
Meitner, Lise, 4
  biographical sketch of, 52–53
  uranium and, 32
Meltdowns, xi, 11, 35, 101
  defined, 228
  economic, 103
  partial, 36
  study on, 37
Mescalero Apaches, storage facility and, 20
Metric ton, defined, 228
Metropolitan Edison v. People vs. Nuclear Energy (1983), 125–126
Michael Mariotte, Nuclear Information and Resource Service (video), 207
Michio Kaku on Nuclear in Space (video), 208
Midland, conversion of, 41
Mill tailings, defined, 228
Milling, 80–81, 84, 233
Millirem, defined, 8, 228
Millstone 1, statistics on, 71
Millstone 2, statistics on, 71
Millstone 3, statistics on, 71
Millstone at the Crossroads (video), 208
Minatom, 173
Mining, 17, 33
  underground/open pit, 80
Minor, G. C., biographical sketch of, 53
Mixed-oxide (mox) fuel, 22, 83
  defined, 228
Mobile Chernobyl, 113–116
Moderator, defined, 228
Molten salt breeder reactor, defined, 228
Monitored retrievable storage, defined, 228
Monticello, statistics on, 71
Mueller, Werner, phase-out and, 27
Multiplication factor, defined, 228
Mutation, defined, 228

Nader, Ralph, 155
  lemon capitalism and, 24
  steam tube rupture and, 44
Nagasaki, 5, 33
Natch 1, statistics on, 68

National Academy of Sciences (NAS)
human intrusion and, 100–101
report by, 89
National Environmental Policy Act (1969), passage of, 37
National Reactor Testing Station (AEC), 34
National Research Council, 22
National Research Experimental Reactor, 34
Native Americans, Yucca Mountain and, 109
Nautilus (submarine), 7
launching of, 34
Neptunium, 98
discovery of, 32
Neptunium-237, 99
Neutron, 3–4
defined, 229
discovery of, 32
poison, 229
slow, 232
New York v. United States (1985), 126
NIMBY (not in my backyard), 21
"Nine Essential Rules of Inquiry" (Gofman), 178
Nine Mile Point 1, statistics on, 71
Nine Mile Point 2, statistics on, 71
Nixon, Richard M.
breeder reactors and, 37
ERDA and, 37
No Nukes Is Good Nukes (video), 208
Nonfossil fuel power, 28
Nonnuclear power, 158
Nonproliferation issues, 105, 152, 173, 174
guide to, 175
North Anna 1, statistics on, 71
North Anna 2, statistics on, 71
Northern Ind. Pub. Serv. Co. v. Walton League (1975), 126–127
NRC. *See* Nuclear Regulatory Commission
Nuclear Age, The (video), 203
Nuclear Awareness News, 187
Nuclear Control Institute (NCI), 140
Nuclear Cover-Up: Chernousenko on Chernobyl—We Shall Die in Silent Ways (video), 212
Nuclear diplomacy, 155
Nuclear energy, 104, 149, 152, 169, 173, 179
defined, 229
development of, xii, 148
discovery of, 1–5
future of, 25–26, 28–29
as percentage of energy supply, 62
See also Nuclear power
Nuclear Energy Agency (U.S. office), 140
Nuclear Energy Info, 187
Nuclear Energy Institute, 140
on safety culture, 9–10
on storage space, 19
Nuclear Energy Research Initiative (NERI), 42
Nuclear engineering, 185
Nuclear Engineering and Design, 187
Nuclear Engineering International, 187
Nuclear Engineering Monthly, 187
Nuclear Europe Worldscan, 187
Nuclear Expansion in Asia and Australia, The (video), 208
Nuclear Index, 187
Nuclear India, 187
Nuclear industry, 160–161, 165, 168, 185
advertisements by, 43
analysis of, 161, 162
Cold War and, 156
commercial, 163
critics/ proponents of, 164
history of, 155, 167
limited liability for, 39
public relations campaigns by, 163
in Sweden, 171
Nuclear Information and Resource Service (NIRS), 141
Nuclear island, defined, 229
Nuclear issues, xiii, 155, 156, 161
interest in, 147
international, 173
overview of/perspective on, 175
public opinion on, 166
understanding, 170
Nuclear Law Bulletin, 187
Nuclear Monitor, 188
Nuclear Nightmare Next Door (video), 213
Nuclear Nonproliferation Act (1978), 120–121
Nuclear pile, 5
defined, 229
Nuclear Plant Journal, 188
Nuclear plants, xii, 155
accidents at, 160
aging of, 26
building, 29
commercial, 7, 172
coal savings and, 28
decommissioning, 23, 25, 41
design of, 59–62

economics of, 154
future for, 23–25
modernizing, 27
operation of, 59–63, 169
opposition to, 24–25
partly finished/inoperable, 167
siting, 165
Nuclear policy, 155
development of, 29, 149, 166, 170
environmental issues and, 37
history of, 162, 163
Nuclear power
accidents and, 101–102
commercial, 166
debate over, 156, 168, 176
development of, xi, 27–28, 31, 105, 153, 154, 162, 164
electricity generated from, 79 (table)
end of, 25–27, 101–105
in Europe, 26–27
in France, 170
future of, xi, 29, 164, 171, 174
global warming and, 104–105
in Japan, xi, 27, 41
legal aspects of, 116–130
oil dependence and, 27
opposition to, 23, 101–105, 152, 155, 158
political history of, 160
problems with, 5–17
proliferation and, 105
in Soviet Union, 177–178
support for, 155, 157–158
technical literature on, 151
using, xii, 165
See also Nuclear energy
Nuclear Power (video), 204
Nuclear Power: Dangerous Energy (video), 209
Nuclear Power: Energy for the Future (video), 210
Nuclear Power Plant Safety: What's the Problem? (video), 204
Nuclear Power Production (video), 204
Nuclear Reaction (video), 204
Nuclear Reactor Safety, 188
Nuclear Regulatory Commission (NRC), 37, 145, 162, 165
casks and, 21
critique of, 159
CSIS and, 161
decommissioning and, 23
establishment of, 11, 38
on meltdowns, 102–103
nuclear power and, 12
nugget file and, 10
plant designs and, 42
regulations by, 26
replacement by, 162
repositories and, 89, 94
roles of, 163
safety violations and, 40
shutdown orders by, 38, 40
Three Mile Island and, 14
Yucca Mountain and, 91, 93, 106
Nuclear Regulatory Reports, 188
Nuclear Resister, 188
Nuclear Safety Research, Development, and Demonstration Act (1980), 39
Nuclear Science and Engineering, 188
Nuclear steam supply system, defined, 229
Nuclear Technology, 188
Nuclear Waste Bulletin, 188
Nuclear Waste Citizens Coalition, 141
Nuclear Waste Fund, 110
Nuclear Waste News, 188
Nuclear Waste Policy Act (NWPA) (1982), 40, 93, 106, 119
disposal issues and, 88
repositories and, 89
temporary waste facilities and, 19
Nuclear Waste Policy Amendment Act (1987), 88
Nuclear Waste Technical Review Board (NWTRB), 88
Yucca Mountain and, 106, 110, 111
Nuclear weapons, xi, xii, 33, 155, 161, 165
capability, 229
defined, 229
development of, 95, 153, 157
dismantling, 22, 168, 181
dropping, 5
materials for, 87–88
nuclear power and, 105
terrorists and, 172
transfer of, 171
Nucleon, defined, 229
Nuclide, defined, 229
Nugget file, 10

Oak Ridge National Laboratory, 18, 180, 181
Occupational Health and Safety Administration (OSHA), 122
Oconee 1, statistics on, 71
Oconee 2, statistics on, 71
Oconee 3, statistics on, 71
Oda, Mayumi, biographical sketch of, 53–54

Official Secrets Act, 171
Ohio Citizens for Responsible Energy v. NRC (1987), 127
Oil crisis, 26, 37, 170
  electricity and, 27
O'Leary, Hazel, 26, 110
Operating costs, defined, 229
Operating license, defined, 229
Oppenheimer, J. Robert
  biographical sketch of, 54
  Manhattan Project and, 33
Organization of Petroleum Exporting Countries (OPEC), 37
Oyster Creek 1, 36
  statistics on, 71

Pacific Gas and Electric Company (PG&E) v. State Energy Resources Conservation and Development Commission (1983), 127
Paducah, 181
  enrichment at, 18
Palisades, statistics on, 71
Palo Verde 1, statistics on, 71
Palo Verde 2, statistics on, 71
Palo Verde 3, statistics on, 72
Parker, James W., 6, 34
Pathway analysis, defined, 229
Peach Bottom 2, statistics on, 72
Peach Bottom 3, statistics on, 72
Peak load, defined, 229
Percolation, defined, 229
Perry 1, statistics on, 72
Photon, defined, 230
Pilgrim 1, 9, 24
  evacuation plans for, 40
  sale of, 42
  statistics on, 72
Plutonium, 105, 115, 165, 172, 173, 182
  controlling, 23
  creation of, 33
  defined, 230
  irradiating, 22–23
  recycling, 22, 83, 84
  reprocessing, 23, 174
  surplus, 23, 43, 87–88
  vitrification of, 181
Plutonium–239, 5, 18, 99
Plutonium–242, 99
Point Beach 1, statistics on, 72
Point Beach 2, statistics on, 72
Pollard, Robert, 12, 159
Pollution, 7
  defined, 230
  thermal, 233
Polonium, discovery of, 32
Portland General Electric, Trojan and, 26
Portsmouth, 181
  enrichment at, 18
Power industry, deregulation of, 25–26
Power plants
  coal burning, 27–28
  defined, 230
Prairie Island 1, statistics on, 72
Prairie Island 2, statistics on, 73
Prairie Island Coalition (video), 209
Precipitators, 27
Pressure vessel, defined, 230
Pressure-tube reactor, defined, 230
Pressurized water reactors (PWRs), 59–60, 230
Price-Anderson Act (1957), 7, 35, 117
  challenging, 39
Primary coolant, defined, 230
Primary loop, defined, 230
Private Ownership of Special Nuclear Materials Act (1964), 36, 117
Progress in Nuclear Energy, 189
Proliferation, 22–23, 42, 149, 161, 173, 174
  defined, 230
  risk of, 182
Proton, defined, 230
Public Citizen Critical Mass Energy Project, 141
Public Citizen Litigation Group, 38–39
Public Citizen, steam tube rupture and, 44
Public policy, defined, 230
Public relations, 163, 164
Public Service Company of New Hampshire, nuclear plants and, 167
Push to Revive Nuclear Power, The (video), 209

Quad Cities 1 and 2, statistics on, 73

Rabi, Isidor I., on nuclear reactions, 3
Rad (radiation absorbed dose), defined, 230
Radiation, 7–9, 170
  alpha, 32
  background, 8, 220
  beta, 32
  defined, 230
  discovery of, xii
  dose rate, 99
  effects of, xi, 8–9, 179, 180
  exposure, 8, 9, 163, 177, 178
  protection against, 179

Radiation (video), 205
Radiation Effects Research Foundation, 142
Radiation Laboratory (Berkeley), plutonium at, 33
Radiation monitors, 16
Radiation sickness, 7, 16, 40, 178
 defined, 231
Radiation therapy, defined, 231
Radioactive materials. *See* Waste
Radioactive Reservation (video), 213
Radioactive tracer, defined, 231
Radioactive Waste Management Association, 142
Radioactivity, 98, 114, 156
 defined, 231
Radioisotopes, 18, 169, 184–185
 defined, 231
Radionuclides, 96, 178, 182
 reducing concentration of, 98–99
 defined, 231
 release of, 89, 97
 water and, 98
Radium, discovery of, 32
Radon, 8
 defined, 231
Radwaste Magazine, 189
Rancho Seco, closing, 25
Ray, Dixy Lee, biographical sketch of, 55
RBMK reactors, 16–17
Reactor Safeguards Committee (AEC), 34
Reactor vessel, defined, 231
Reactor year, defined, 231
Reactors, xii
 canceling, 26
 defined, 231
 electricity from, 79
 first, 4–5, 33, 169
 fuel for, 22
 inspection of, 11
 licensing, 11, 24
 North American, 63 (fig.)
 poisoned, 15
 shutdown of, 15, 24, 104
 U.S., 63, 63–77 (table)
 uses of, 181
 world, 77, 77 (fig.), 78 (fig.)
Reagan, Ronald, Nuclear Power Policy Statement and, 39
Recycling, 83, 84
 defined, 231
Reference design, 91–95
Refueling, defined, 231
Regulation, 17, 116–119, 166, 167, 176
Regulatory agencies, 121–122
Regulatory policy, 163
 defined, 231
Rem (radiation equivalent man), defined, 8, 231
Renewal energy sources, defined, 231
Repair enzymes, defined, 232
Report to the President (Interagency Review Group on Nuclear Waste Management), 90
Repositories, 89, 181, 187
 closing, 94–95, 99
 cooling of, 96
 defined, 232
 design for, 88, 90, 91–95
 locating, 19
 safety issues at, 99–101
 temporary, 19
 water and, 107
 See also Storage
Reprocessing, 22, 23, 27, 80, 83, 174
 defined, 232
Research
 history of, 172
 nonmilitary, 39
Riccio, Jim, 24
Rickover, Hyman G.
 biographical sketch of, 55–56
 Nautilus and, 7, 34
River Bend 1, statistics on, 73
Robert E. Ginna, statistics on, 73
Roberts v. Florida Power and Light (1998), 127–128
Rock salt formations, 89
Rocky Mountain Institute (RMI), 104
Roentgen, defined, 232
Roentgen, Wilhelm, 1, 31
Roosevelt, Franklin D.
 atomic bomb and, 33
 Einstein and, 5, 33
 Manhattan Project and, 33, 160
Royal Institute of International Affairs, 173
Rutherford, Ernest, 3, 32

Safeguards, defined, 232
Safety culture, 9–10
Safety issues, 39, 40, 152, 163, 164, 172, 176, 179, 183, 185
 dealing with, 9–12, 17
 repository, 99–101
Safety system, defined, 232
Safstor, 23–24, 232
St. Lucie 1, statistics on, 74
St. Lucie 2, statistics on, 74
Salem 1, statistics on, 73

Salem 2, statistics on, 73
San Onofre 1, closing, 25
San Onofre 2, statistics on, 73
San Onofre 3, statistics on, 73
Savannah River, 20, 108
Schroeder, Gerhard, 44
Scram, defined, 232
Scranton, William W., III, 153
Scrubbers, 27
Seaborg, Glenn T., biographical sketch of, 56
Seabrook 1, 167
  controversy over, 162
  statistics on, 74
Seabrook: Do We Need It? (video), 205
Secondary loop, defined, 232
Security issues, 172
Sellafield nuclear processing plant, 171
Sequoyah 1, statistics on, 74
Sequoyah 2, statistics on, 74
Shearon Harris 1, statistics on, 74
Shielding, defined, 232
Shippingport nuclear power station, 7, 34
  decommissioning of, 25, 39
  history of, 35, 166
Shoreham nuclear power plant, 166
  closing, 25
  controversy over, 162
Shutdowns, 15, 24, 38, 104
  defined, 232
  report on, 40
Sierra Club, nuclear power and, 7
Silkwood, Karen, biographical sketch of, 56–57
Silkwood v. Kerr-McGee Corporation (1984), 128, 180
Site characterization, poor management of, 109–110
Sixty Minutes to Meltdown (video), 212
Slovik, Paul, 25
Social Democrat-Green party coalition, nuclear power and, 43
Soddy, Frederick, 3, 32
Software, commercial nuclear reactors and, 184
Solar power, 28, 154
  space program and, 184
South Texas 1, statistics on, 74
South Texas 2, statistics on, 74
Space program, nuclear power for, 184
Spent fuel, 21, 38, 82, 98, 114
  commercial, 87
  defined, 232
  disposing of, 83–84
  locating storage of, 85
  pool, 233
  reprocessing, 23
  storing, 18–19, 83, 86 (fig.)
  total inventory of, 88
Stable nucleus, defined, 233
STAR (Standing for Truth About Radiation), 142
State of Wisconsin v. Northern States Power Company (1985), 128
Statute of the International Atomic Energy Agency (IAEA) (1957), 119
Steam generators
  defined, 233
  replacing, 26
  rupture of, 44
Storage
  defined, 233
  locating, 85, 86 (fig.)
  temporary, 41, 85, 227
  See also Repositories
Strassmann, Fritz, 4
  barium/krypton and, 32
Strauss, Lewis L., 7
  biographical sketch of, 57
Strontium–90, problems with, 18
Summer 1, statistics on, 75
Surry 1, statistics on, 75
Surry 2, statistics on, 75
Surry nuclear plant, accident at, 40
Susquehanna 1, statistics on, 75
Susquehanna 2, statistics on, 75
Swedish Energy and Environmental Policy (SEEP) Model, 171
Szymanski, Jerry, on earthquakes/Yucca Mountain, 107

Tarapur atomic power station, start up for, 36
Technetium, 98
Technetium–99, 99
Technology
  controversies surrounding, 29
  separation, 182
Tennessee Valley Authority, Watts Bar reactor and, 103–104
Terrorists, 22
  nuclear weapons and, 172
  uranium/plutonium and, 18
"Theoretical Possibilities and Consequences of Major Accidents in Large Nuclear Plants," release of, 35
Theory of relativity, 32
Thermal reactor, defined, 233
Thompson, Joseph John, 2, 31

Thorium–232, defined, 233
Thorpe, Grace, biographical sketch of, 57–58
Three Mile Island, 24, 25, 40, 147, 161, 169
accident at, xi, 12–14, 39, 102
clean up at, 14
health damages from, 43
investigating, 151, 153
legacy of, 14, 17, 23, 157, 176
sale of, 42
study on, 41, 44
Three Mile Island 1, statistics on, 75
Three Mile Island Revisited (video), 212
Time Out for Science: Benefits and Uses of Nuclear Energy (video), 210
Tokamak, 41
defined, 233
Topical Meeting on Artificial Intelligence and Other Computer Applications, 185
Toshiba, ABWR by, 41
Train v. Colorado Pub. Int. Research Group (1976), 128–129
Transmutation, 32, 182
Transportation issue, 20–22, 41, 98–99, 113–114, 115, 179
Transportation of Nuclear Materials, The (video), 210
Treaties, international, 119–121
Treaty of Ruby Valley (1863), 109
Treaty on the Non-Proliferation of Nuclear Weapons (1968, 1970), 120, 175
Triple play reactors, 23
Tritium, 23, 105
Trojan, closing, 25, 26
Truman, Harry S, Atomic Energy Act and, 6, 33
Turbine, defined, 233
Turkey Point 3, statistics on, 75
Turkey Point 4, statistics on, 75

Udall, Stewart, biographical sketch of, 58
Underground facilities/operations, 43, 92–93
Union of Concerned Scientists, 38, 143
nugget file and, 10
study by, 42–43, 159
Union of Concerned Scientists v. NRC (1987), 129
United Nations, IAEA and, 35
United Nations International Conference on Peaceful Uses of Atomic Energy, 35
United States Department of Energy Office of International Nuclear Safety and Cooperation, 144
United States Department of Energy Office of Science, 144
United States Department of Energy Office of Scientific and Technical Information, 144
United States Environmental Assessment of a Yucca Mountain Repository (DOE), 115
United States of America and Trustees of Columbia University v. City of New York (1978), 129
University of Pittsburgh, Three Mile Island study by, 44
Uranium, 2, 32, 79, 98, 115
concentration of, 80
defined, 233
depleted, 84, 223
enriched, 18, 181, 183, 226
milling, 80–81, 233
mining, 17, 33, 80
recycling, 83, 84
shipping, 18
weapons-grade, 39, 234
X-rays and, 31
Uranium (video), 214
Uranium dioxide, 81
Uranium hexafluoride, 18, 81, 82, 182
Uranium Information Centre (Australia), 145
Uranium Institute, The, 145
Uranium Mill Tailings Radiation Control Act (1978), 118
Uranium oxide, 82
Uranium-233, defined, 233
Uranium-234, 99
Uranium-235, 5, 18, 22, 35, 81, 82, 84
defined, 233
reprocessing and, 83
Uranium-238, 5, 18, 81, 82, 84
defined, 233
reprocessing and, 83
U.S. Geological Survey, 22
disposal and, 91
hydrological investigations by, 107
Yucca Mountain and, 106
Usenet newsgroups, xiii

Vermont Yankee 1, statistics on, 76
Vermont Yankee Power Corp. v. NRDC (1978), 129–130
Villard, Paul, gamma rays and, 32
Vitrification, 108, 181

defined, 234
Vogtle 1, statistics on, 76
Vogtle 2, statistics on, 76
Volcanism
waste burial and, 100
Yucca Mountain and, 107–108

War and Peace in the Nuclear Age (PBS series), 155
Ward Valley, waste site in, 21–22
Wasco, 165
Washington International Energy Group, 104
Washington Nuclear 2, statistics on, 76
Waste, 173, 178, 184
American West and, 181
containing, 84, 87
controversy over, 105–112, 150
defense, 108
defined, 231
disposing of, xiii, 80, 91, 154, 181, 183
laws/regulations about, 88–89
long-lived, 104
management of, 79, 168, 179, 180
medium-level, 84
problems with, xii, 17–22, 85, 87
spread of, xiii, 182
storing, 41, 169, 183
transporting, 20–22, 41, 179
transuranic, 233
water and, 97–98
See also High-level waste; Low-level waste
Waste disposal system, defined, 234
Waste facilities. *See* Repositories
Waste Isolation Pilot Project (WIPP), 42
Waste packages, 90, 94
cooling of, 96
copper, 89
criticality and, 101
designs for, 93
evaluating, 97
lifetime of, 97
limiting water contact with, 96–97
shipping, 98–99
Water
repositories and, 107
waste and, 97–98
Water table, defined, 234
Waterford 3, statistics on, 76
Watkins, James, 109
Watt, defined, 234
Watt-hour, defined, 234
Watts Bar 1, 24, 103–104
statistics on, 76
Weiss, Ellyn, 159
Wells, H. G., atomic energy and, 3, 32
Western Shoshone, Yucca Mountain and, 109, 115
Western States Legal Foundation, 146
Westinghouse
nuclear market and, 36
reactor sales by, 105
What'll We Do with the Waste When We're Through? (video), 214
Wilson, Carroll, AEC and, 11
WISE (World Information Service on Energy), 146
Wolf Creek, statistics on, 76
Wolfson, Richard, on risks, 9
World Association of Nuclear Operations, 146
World Nuclear Industry Handbook, 189
World Set Free, The (Wells), 32
World Wide Web, nuclear issues and, xiii

X-rays, 1, 2, 31

Yankee Nuclear Power Station, 36
Yankee Rowe, closing, 25
Yellowcake, 18, 81, 234
Yucca Mountain, 88, 182
characterization of, 109–110
choosing, 90–91
conditions at, 19–20, 90, 100, 106
confirmation/retrieval at, 94
delays for, 43
disposal at, 44, 85, 87, 106–112
independent review of, 111–112
opposition to, 105–112
reference design for, 91–95, 92 (fig.)
surface facilities/operations at, 91–92
underground facilities/operations at, 92–93
waste shipments to, 113
water at, 96, 107

Zion 1, statistics on, 76
Zion 2, statistics on, 77
Zircaloy, 93
Zirconium tubes, 18